Reading
the Clouds

HOW YOU CAN FORECAST THE WEATHER

Oliver Perkins

ADLARD COLES

LONDON · OXFORD · NEW YORK · NEW DELHI · SYDNEY

ADLARD COLES
Bloomsbury Publishing Plc
50 Bedford Square, London, WC1B 3DP, UK
29 Earlsfort Terrace, Dublin 2, Ireland

BLOOMSBURY, ADLARD COLES and the Adlard Coles logo are trademarks of
Bloomsbury Publishing Plc

First published in 2017 as *The Message of the Clouds*
Expanded edition published as *Reading the Clouds* by Adlard Coles in 2018
. This edition published 2023

For legal purposes the Acknowledgements on p. 119
constitute an extension of this copyright page

A catalogue record for this book is available from the British Library

Library of Congress Cataloguing-in-Publication data has been applied for

ISBN: PB: 978-1-3994-0142-5; ePub: 978-1-3994-0141-8; ePDF: 978-1-3994-0140-1

2 4 6 8 10 9 7 5 3

Typeset in Freight Text Pro by Carr Design Studio
Printed and bound in India by Replika Press Pvt. Ltd.

To find out more about our authors and books visit www.bloomsbury.com
and sign up for our newsletters

This book is dedicated to Jesus Christ

The heavens declare the glory of God;
The skies proclaim the work of his hands

Psalm 19:1

CONTENTS

FOREWORD
BY TOM CUNLIFFE

It's so much better, and far more fun, to be able to read the sky and make your own conclusions about the weather than it is to punch a button on a computer and let someone else tell you what's happening. Of course, we're all going to use modern methods, but it's like anything else, if you understand the essence of the subject, you know what questions to ask the computer and can make a lot more sense of its answers. That's what Olly's book does. It leads you gently through the guts of what I used to call single-station weather forecasting.

I've known Olly since he was a small boy. I was impressed then by his keenness and the intelligent questions he asked. Now he's a strapping young teenager, he's moved on and is already representing his country in sailing. His first book is so good that my Yachtmaster Instructor candidates would do well to read it. Driven by instantly available internet forecasting, the subject of weather lore is falling off most people's radar. This is a poor state of affairs because, out at sea, there may only be the sailor to decide on what's happening and what to do about it.

Olly has given us all a helping hand. I learned something from this book. I'll bet you do too.

Tom Cunliffe

Author, TV presenter and Yachtmaster Instructor Examiner
www.tomcunliffe.com

FOREWORD

BY DUNCAN WELLS

I met Olly when we were both on the ChartCo Stand at the 2017 Southampton Boat Show, signing copies of our books. I love weather, especially simplifying it so that my students can understand it more readily. I am also a great fan of Alan Watts' Instant Weather Forecasting. And let's face it, if that book was good enough for Bernard Moitessier to refer to when he was unsure of the weather during the 1968 single-handed Golden Globe race then it's good enough for me. So I was intrigued at what Olly was offering. I bought a copy of his original book and found it absolutely brilliant; a must for anyone who does anything outside and for whom the weather might be important. It's for sailors of course but gardeners, walkers, golfers – everyone really, wherever they are in the world, will get something from this book.

And so while I thought that it was admirable to have a self-published book, as it was at the time, I thought it would be better to have a big publishing name behind it. I contacted my publisher, Janet Murphy at Bloomsbury, and here is the result. With *Reading the Clouds*, Olly has expanded the original content and added more photos, diagrams, tables and explanations, as well as more tips and hints on identifying and interpreting cloud formations.

Thanks to Olly you will be able to look at the sky mid-morning and tell with certainty what the wind will be doing and whether or not it will rain during the afternoon. That's worth knowing.

Duncan Wells

Principal of Westview Sailing, creator of MOB Lifesavers
and author of the *Stress-Free* series of sailing books

www.westviewsailing.co.uk

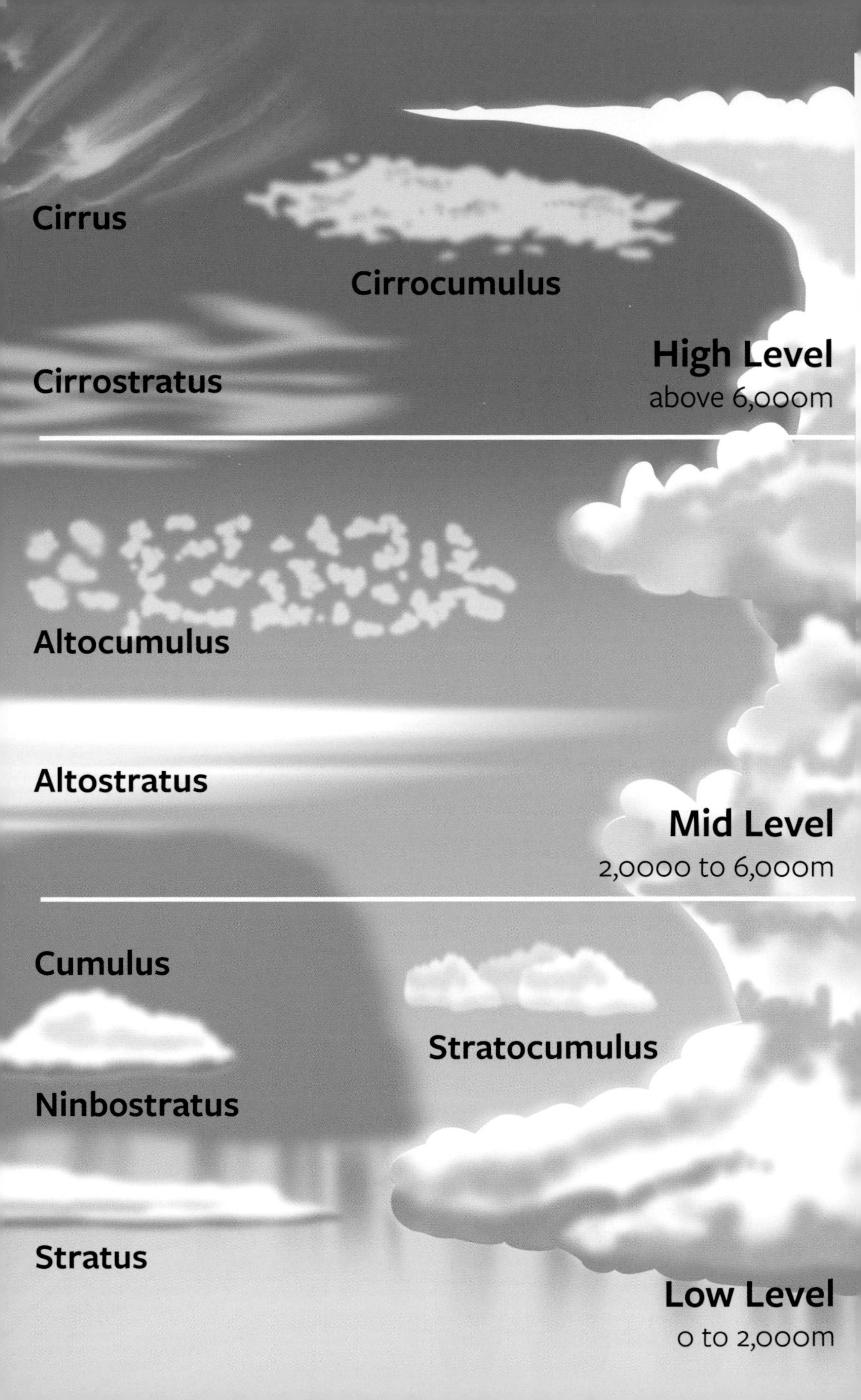

Cirrus

Cirrocumulus

High Level
above 6,000m

Cirrostratus

Altocumulus

Altostratus

Mid Level
2,0000 to 6,000m

Cumulus

Stratocumulus

Ninbostratus

Stratus

Low Level
0 to 2,000m

GETTING TO KNOW THE BASICS

1

This chapter will explain the basics of weather so that you can understand why clouds indicate different types of weather. It covers clouds, air masses, depressions, fronts, anticyclones and some forecasting rules. If you are not particularly interested in the theory behind how the weather works but want to start predicting what will happen as soon as possible, skip to the section on the cross-winds rule on page 19.

CLOUDS

Clouds are formed when air can no longer carry any more water vapour, so the vapour condenses into droplets. This often happens due to warmer air rising as it is less dense than cooler air. Rising air cools at approximately 1°C per 100m of altitude gained. Since colder air cannot hold as much water vapour as warm air, there is an altitude when the water vapour condenses into clouds. However, at some point the air stops rising as it reaches a level where the air gets warmer the higher it goes. This is called a temperature inversion, and there is a permanent one called the tropopause which is about 12km high in the temperate regions.

The tropopause acts as a lid in the atmosphere, so almost all clouds are contained in the troposphere which is the layer between the ground and the tropopause. The clouds are split

◄◄ The different types of cloud

into three layers. The lowest layers are the clouds less than 2km above the ground and their names do not have a prefix. The next layer contains clouds up to 5km high and they all start with the prefix 'alto-'. The highest layer is made of ice crystals and is over 5km high, they start with prefix 'cirr-'. There are also clouds which extend through more than one layer.

The low clouds are: cumulus, the iconic cotton wool cloud; stratus, the featureless cloud that covers the whole sky; and stratocumulus clouds, which are a mix of the two.

The middle clouds are: altocumulus clouds, which look like lots of smaller cumulus clouds; and altostratus, which are just higher stratus clouds.

The high clouds are: cirrus clouds, the thin wispy clouds; cirrostratus clouds, thin layers of ice crystals that can form a halo round the sun; and cirrocumulus clouds, which are higher versions of altocumulus clouds.

The clouds that span more than one layer all produce rain. They are nimbostratus, which is the dull rain-bearing cloud, and cumulonimbus, which is the towering shower and storm cloud.

AIR MASSES

Air masses are bodies of air in which the temperature and humidity are constant. They are separated from each other by a sharply defined transition zone, called a front. These air masses are usually hundreds or even thousands of miles across. They get their characteristics for staying over one area for weeks or even months. This is called a source region, though air masses often end up moving away from their source region.

There are four main types of air masses in the temperate regions: maritime polar, maritime tropical, continental polar

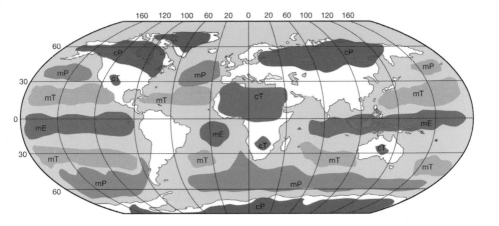

and continental tropical. See the diagram for their locations around the world.

▲ Locations of air masses

Maritime polar (mP) air comes to the UK with a north-westerly wind behind cold fronts from its source region near Iceland or over the North Pacific. This maritime polar air is cold, clean, humid and has excellent visibility. Also, it often brings showers in the afternoon, or over the sea at any time of the day.

Maritime tropical (mT) air comes with a south-westerly wind from its source region over the Sargasso Sea or over the Central Pacific. This air is warm and very moist. With maritime tropical air, there are lots of low clouds and there is often drizzle or rain over the land. There can also be fog on coasts and over hills. The visibility is poor. It is maritime tropical and maritime polar air which lead to rainy frontal weather.

Continental polar (cP) air comes with an easterly or north-easterly wind from Northern Europe or Canada. This often leads to cold winter days, but it rarely brings rain. This air is quite rare in the UK.

Continental tropical (cT) air comes with a south-easterly wind from its source in North Africa or around Colorado.

This provides heat waves in the summer and drops Saharan dust in Europe. This air is also rare in the UK.

DEPRESSIONS

Depressions are areas of low air pressure which bring rain, clouds and strong winds because of rising air inside them. They are experienced throughout the year in the temperate climates and they travel around the globe from west to east. Britain's depressions form over the Atlantic Ocean when warm, moist maritime tropical air from the south meets colder, drier and denser maritime polar air from the north. The two air masses do not mix as they have different temperatures and densities. So, a front is formed between these two air masses. When the warm air moves towards the denser and colder air it is forced to rise above the cooler air; this is called a warm front. When the cold air moves towards warm air and undercuts the warm air, it is known as a cold front.

The depressions described below are the depressions when they are at their most ferocious. Most depressions will be in a stage of decay by the time they reach the eastern parts of the UK and most of Europe, and they may have begun to occlude. Occluding is where the cold front catches up with the warm front and undercuts the warm air completely (see page 18).

Warm front or occlusion approaching

Fronts always slope at a shallow angle – less than one degree for a warm front and about two degrees for a cold front. This means that we can foretell a warm front due to the cloud sequence which forms over 1,000km ahead of its arrival on the ground.

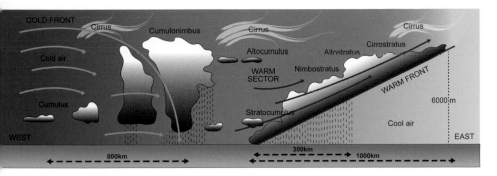

Although they are not always indicative of a front, the first signs of a warm front are thick aircraft trails (contrails). These form due to damp air. The most important cloud to memorise is cirrus. Cirrus clouds are high, thin and wispy. If the number of cirrus clouds in the sky is increasing it is a good indication that a depression is coming in around 15 hours. Often there are cirrocumulus clouds (high lumpy clouds) a couple of hours after the cirrus clouds are first seen. However, cirrocumulus clouds aren't always a sign of an approaching front.

As the warm front nears, the clouds get lower and thicker. Cirrostratus clouds give the sun a halo and suggest that the front is about 12 hours away. Then lower altostratus clouds uniformly cover the sky. These indicate that rain is just a few hours away. As the altostratus thickens, the sun will get dimmer and dimmer and eventually the sun will no longer be seen through the cloud.

Once the rain starts to fall, the front is about 4 hours away. As the front draws nearer, the rain falls more strongly, the winds increase and the cloud base lowers. This rainy stratus cloud is called nimbostratus, meaning 'rain-bearing layer'. Throughout the approach of a warm front, the wind will back (shift anti-clockwise) a little as well as increase considerably.

The speed at which fronts pass is difficult to determine, so the above timings can vary significantly. However, there is a

▲ Cross section of a depression

weather proverb which states 'Long foretold, long last. Short notice, soon past'. This is often very accurate because if the cloud build-up is quick, the depression will also pass quickly.

Learn this sequence of clouds ahead of warm and occluded fronts:

1. **Cirrus** – the wispy clouds
2. **Cirrostratus** – halos around the sun or the moon
3. **Altostratus** – starts to obscure the sun
4. **Nimbostratus** – rain!

Warm sector

As the warm front passes, there is a quick rise in temperature and the wind veers (shifts clockwise). The warm sector has arrived. Throughout this book, in the southern hemisphere, read a back (shift anticlockwise) as a veer and read a veer

▼ A lowering cloud base before a warm front

▲ Warm sector weather

as a back, as the winds go the opposite way around weather systems. It can be difficult to know when the warm front is passing as there is not such a clear-cut change as that of a cold front. The first sign of the warm front passing is that the cloud base starts to lift, although there is sometimes a low ragged band of dark cloud marking the front. In the warm sector there is low and broken cloud, decreasing wind from a constant direction, and the rain often stops. The warm sector usually has poor visibility. Also, you will often see wisps of low stratus clouds.

There will often be short sunny spells, but there should always be more cloud than sky. There may also be drizzle on windward coasts or slopes. Sometimes the warm sector can last for days; other times it is gone in hours. There is no way of knowing which it will be without a weather chart so the rain could last for days, especially on slopes. However, the further south you are the larger the warm sector usually is.

Cold front passing

Cold fronts are much more ferocious than warm fronts and it is harder to predict when they are going to arrive. When the cold front passes through, winds can often reach gale force and the rainfall is very heavy, although it does not often last long. The air temperature will also fall rapidly and the wind will veer considerably. The wind veer can often come quite suddenly and will be obvious to sailors. This vigorous weather passes quickly and leaves much lighter rain and often altostratus clouds. The main sign that a cold front is approaching is the squall that typically blows through as it arrives.

▶▶ Cloud sequence after a cold front

Accompanying a cold front will be cumulonimbus clouds, which are tall convectional clouds that produce heavy rain, snow or hail. These cumulonimbus clouds will be difficult to see as there will often be clouds in front of them obscuring their size until they are overhead.

Once the cold front passes there are usually cumulus and altocumulus clouds. Then there will be cumulonimbus clouds and cumulus congestus clouds (cumulus clouds that are taller than they are wide). Next some altostratus clouds will form, then cirrus clouds will be left behind and finally even the cirrus clouds will pass. This whole process takes just a matter of hours.

These three photos show the different clouds that are seen as a cold front passes. The first image is under a cumulonimbus cloud where it was raining heavily and the cold front was directly overhead. The next photo was taken 45 minutes later and it shows smaller cumulus clouds; at this

point there was no longer any rain. The final photo was taken another 45 minutes later and it shows fewer cumulus clouds, but cirrus clouds can be seen through the gaps in the cumulus clouds.

Occluded fronts

Occluded fronts are fronts where the cold front has caught up with the warm front and undercuts the warm air completely. Fronts occlude at the end of their lives and they have characteristics of both warm and cold fronts, but are much less active. They often bring showers and light rain. Drizzle is prevalent as it is formed when the warm air is forced up above the cold air and then must shed water as it cools. There is not the poor visibility of the warm sector and the wind is much lighter than that of a cold front.

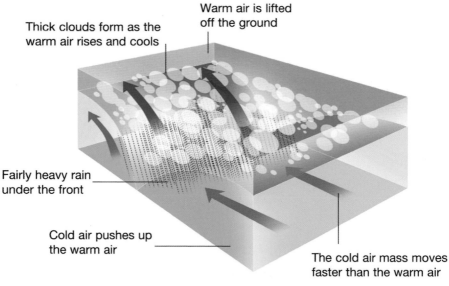

Warm air is lifted off the ground

Thick clouds form as the warm air rises and cools

Fairly heavy rain under the front

Cold air pushes up the warm air

The cold air mass moves faster than the warm air

▲ Cross section of an occluded front

ANTICYCLONES

Anticyclones are the opposite of depressions; they are areas with higher pressure than the surrounding area, so the air descends in the middle. This descending air suppresses cloud formation, which often leads to an area of settled weather. In the summer anticyclonic conditions bring hot and dry weather. However, in the winter anticyclonic conditions are colder with clear skies and calm conditions, which lead to very cold nights.

Anticyclones often cause temperature inversions, where the air higher up is warmer than the air below it. These conditions are very stable and can lead to sheets of widespread stratiform clouds which can persist for days even in the summer. This weather is often known as 'anticyclonic gloom'.

Anticyclones often stay in the same location for a long time and they also have a tendency to block the path of depressions. These are called blocking highs. There are often blocking highs over Spain and the Mediterranean which is why they have so few weather systems. If these blocking highs move to other areas, they can cause unseasonal weather in some areas. Blocking highs over Scandinavia mean that weather systems are deflected south towards the UK, which is when Britain gets some of its worst weather.

THE CROSS-WINDS RULE

The upper wind generally travels from west to east, along the path of the depression, and the lower wind moves around a depression anticlockwise. Anyone can use this knowledge to work out whether they are in the path of a depression.

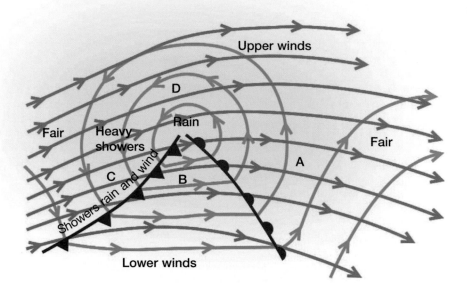

Upper winds

D

Rain

Fair

Heavy
showers

Showers rain and wind

C

B

A

Fair

Lower winds

▲ Upper and lower
winds around a
depression

The diagram on the next page shows a depression with the isobars of the upper winds (in green) and the lower winds (in blue). An isobar shows areas of equal air pressure and the closer they are together, the windier it will be. The isobars also roughly mark the wind direction. At position A the lower wind is at right angles to the upper wind. The wind is also at right angles at C. As depressions move from the west to the east, at position A the weather will deteriorate and at position C the weather will improve. At position B the lower wind will be going in the same direction as the upper wind and a cold front will arrive. Also at position D, the winds will be going in opposite directions and the weather will stay the same. From this, the weather in the next few hours can be predicted. Therefore, we have these rules:

- If you stand with your back to the lower wind and the upper wind moves from left to right the weather will get worse.
- If you stand with your back to the lower wind and the upper wind moves from right to left, the weather will improve.

- If the lower wind is going with or against the upper wind, then the weather is unlikely to change much, unless you are in the warm sector.

The first two rules only apply in the Northern Hemisphere. To make them work for the Southern Hemisphere stand facing the wind instead.

> **Quick rule:** If that all sounded too complicated just remember this. If the winds are perpendicular then the weather will get worse, unless you have just had a depression pass. If the winds are parallel to each other then there should be fair weather.

Finding the wind direction

Finding the direction of the lower wind isn't as simple as it seems. It is important to note that due to friction the surface wind is backed (anticlockwise) compared to the gradient wind. The angle of this back is about 15 degrees over the sea and 30–40 degrees over the land. The wind tries to flow from areas of high pressure to areas of low pressure but due to the rotation of the earth it flows around a depression like water around a plug hole. When the wind is in contact with the ground there is more friction, so it slows and instead of being forced to spiral by the earth's rotation it heads closer towards areas of low pressure. Because of this, it can be easier to use the passage of the lower clouds, such as cumulus clouds to work out the direction of the lower wind.

Working out the upper wind direction can be quite difficult, but there are a number of ways of working it out. One way is to use a fixed point, such as a tree, or to use a celestial body such as the sun or some stars. It should then be possible to see the upper clouds moving past that point over the period of a few minutes. Another way is to use the fallstreaks of cirrus clouds. Fallstreaks are where the cirrus clouds have snow falling from the bottom of them, which evaporates well before it reaches the ground. As the fallstreaks fall they slow down in the calmer lower winds. This means that they will always trail behind the cirrus clouds, pointing the direction of the upper wind. For example the upper wind in the photo on the next page is going from the top right to the bottom left.

If there are persistent contrails (see page 32), then these can also help you to find the upper wind direction. These contrails slowly dissipate and if the wind is coming perpendicular to their direction you will notice that they will shred sideways, but if the contrails seem to grow little turrets, then it indicates that the wind is going along them. These, however, don't reveal exactly which direction it is going in, only whether it is going backwards or forwards across or along the contrail.

In the photo on page 33 the upper wind appears to be going across the page, but it is difficult to work out whether it is coming from the left or right.

▼ Fall streaks falling from cirrus clouds

READING THE HIGH CLOUDS

2

The high clouds are extremely useful as the first sign of a warm or occluded front approaching. They are all frozen clouds and they do not produce any precipitation.

▲ Numerous cirrus clouds indicate that a warm or occluded front is coming in 12-24 hours

▲ Cirrocumulus clouds indicate that a warm or occluded front is coming in around 10–15 hours

► A halo around the sun or moon indicates that a warm or occluded front is coming in around 10–15 hours

▲ Abnormally thick condensation trails can indicate a warm or occluded front in 15–30 hours

▲ Cirrus clouds after a cold front can indicate fair weather in a couple of hours

CIRRUS CLOUDS

Cirrus clouds indicate that a warm or occluded front is coming in 12–24 hours. Cirrus clouds are ice-crystal clouds and they are the highest of the clouds in the troposphere. Cirrus clouds often have fallstreaks, which are ice crystals falling from the clouds. These ice crystals slow down as they fall into the lower and slower winds, so they often trail behind the cirrus clouds. Cirrus clouds are usually a wispy shape, either in straight lines, hooks, or in dense tangles.

If cirrus clouds start increasing in number, it often means that a warm or occluded front is approaching. Other signs include the temperature increasing and the barometer falling. In addition, if you start seeing other clouds such as cirrostratus, cirrocumulus or altostratus then you can be confident a front is coming.

By using the cross-winds rule you can confirm whether a depression is approaching or not.

▼ Cirrus clouds before a depression

HOOKED CIRRUS CLOUDS

Hooked cirrus clouds indicate that the warm front of a vigorous depression is approaching in 12–20 hours. If the fallstreaks from cirrus clouds are hooked, then it suggests that there are different wind directions at different levels. This is called wind shear and it often precedes a depression. The sharper the hook, the quicker the wind direction changes at altitude and the stronger the upper wind. Strong upper winds and quick wind direction changes suggest that a depression is approaching, that the wind will reach gale force and that precipitation is coming.

Again, by using the cross-winds rule you can confirm whether a depression is approaching or not.

▼ Hooked cirrus clouds seen about 15 hours before the warm front of a vigorous depression

JET STREAM CIRRUS

►Jet stream cirrus clouds – 50 knot winds were recorded after these clouds were seen

Jet stream cirrus clouds indicate that winds will increase significantly in 6–12 hours. Cirrus clouds often suggest that a warm or occluded front is coming. However, jet stream cirrus means that an especially vigorous depression is on its way.

Jet stream cirrus is formed by the fallstreaks of cirrus clouds grouping together in fast high altitude winds called the jet streams. The jet streams have a wind speed of about 100 knots (a knot is about 1.2mph) so jet stream cirrus can often be seen moving across the sky; this is seen particularly clearly when reference points like a tree or the sun, moon and stars are used. These form long streaks or banners that converge towards the horizon, though they will only form if there is lots of moisture in the air. They suggest that the wind will reach gale force and that precipitation is coming. You can use the cross-winds rule to confirm whether these indicate a front. However, if these clouds can be seen it is very likely a warm front is coming. Jet stream cirrus clouds suggest that a front will pass through in about 12–20 hours.

CIRRUS INDICATING IMPROVEMENT

Cirrus clouds after poor weather indicate improvement within 1–2 hours. Cirrus clouds also appear after cold fronts, but they are only visible for a short period, after which there will be a clear sky with some cumulus clouds. If you see cumulonimbus clouds, then cirrus clouds and large cumulus clouds, improvement is very likely. The cross-winds rule will confirm that another front is not coming. These clouds indicate that there will be no more frontal activity and there will be clear skies with cumulus clouds during the day. They also suggest that within a day the wind will drop to normal levels.

▼ Cirrus clouds indicating improvement

FAIR WEATHER CIRRUS

A few cirrus clouds with parallel upper and lower winds indicate fair weather. Cirrus clouds often indicate fronts. However, when the cross-winds rule suggests that no fronts are coming it means that the cirrus clouds are indicating fair weather. Fair weather cirrus clouds do not increase dramatically in number. They are often arranged in a few separate shreds of cloud, but if these clouds build up then poor weather can be expected. Use a barometer if you have one to hand and the pressure is stable then this is almost certainly fair weather cirrus. These clouds suggest light to moderate winds and often no cumulus clouds. If the lower wind starts to back and the pressure starts to drop then be careful as a front may now be coming.

▼ Fair weather cirrus

CIRROSTRATUS CLOUDS

Cirrostratus clouds indicate that a warm or occluded front is coming in 10–15 hours. Cirrostratus clouds are sheet-like and they often cover the whole sky. Though it is always possible to see the sun or the moon through them, a halo often forms around the sun or the moon, confirming that the cloud is cirrostratus.

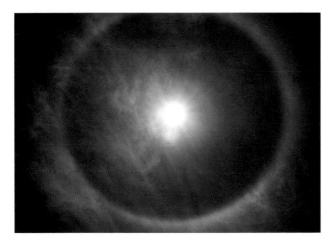

◀ A halo in cirrostratus cloud

Cirrostratus clouds often form ahead of warm fronts and will gradually thicken as the front draws closer. This sheet of cloud slowly lowers until it becomes altostratus cloud. Cirrus clouds are also often present with cirrostratus clouds, and seeing both cirrus and cirrostratus clouds means it is very likely that a warm front is coming. They also suggest that the wind will build to a near gale and precipitation will come later.

Again, using the cross-winds rule can confirm whether a depression is approaching or not. These cirrostratus clouds suggest that a front will arrive in 10–15 hours.

CIRROCUMULUS CLOUDS

Cirrocumulus clouds can indicate that a warm or occluded front is coming in around 10–15 hours. Cirrocumulus clouds are made up of tiny regular spaced cloudlets that do not have any shading. The sun can also shine through the cloudlets. These are known as mackerel skies as they look like fish scales, they often precede a warm front or occluded front. However, they do not always indicate a front, so it is best to look out for other clouds or the crosswinds rule too. If cirrocumulus clouds do indicate a front, the front should arrive in around 10–15 hours.

▶ Cirrocumulus clouds

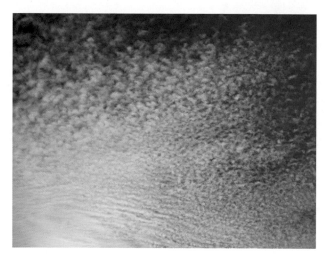

CONTRAILS

Abnormally thick condensation trails (contrails) can indicate a warm or occluded front arriving in 15–30 hours. Contrails are formed by water vapour condensing around aircraft exhaust particles and they will only form if the air is moist.

One of the times when the air is moist is before a front. The longer the contrail lasts for, the more humid the air. Contrails last for between 10 minutes to a few hours. Contrails also form in anticyclones, although they are generally thinner and less long lasting, so their presence doesn't necessarily mean a front is arriving. This means to confirm a front's arrival you must use the crosswinds rule and other cirrus clouds.

Did you know?

Although contrails look like harmless clouds they actually contribute significantly to global warming. This is because they don't reflect much sunlight away from the earth, but trap the heat radiating out from the earth at night. This is different to lower clouds, such as cumulus or stratus clouds, which reflect more of the sun's heat than they trap.

▼ Contrails before a warm front

READING THE MIDDLE CLOUDS

3

The middle clouds are helpful for confirming whether a warm or occluded front is coming after cirrus clouds have been seen. They are, however, not very useful for much other weather forecasting, although they can indicate an unstable atmosphere leading to thunder later. The middle clouds are a mix of ice and water.

ALTOSTRATUS INDICATING A FRONT IS COMING

A lowering cloud base with altostratus clouds indicates that a warm or occluded front is coming in around 8 hours. Altostratus clouds are relatively featureless, but the sun can often be seen through them. On their own, they do not always indicate a front is coming. However, if cirrus clouds were observed before the altostratus arrived it usually means that a warm or occluded front is approaching. Also, if the altostratus is arriving and lowering from the west, it suggests that a front is coming even if the cirrus clouds were missed earlier.

These altostratus clouds will gradually thicken and lower into rain-bearing nimbostratus clouds. Seeing these clouds suggests that the wind could increase up to gale force and that rain is only a few hours away. As quite a rudimentary method of telling whether it is a depression or not, bear in

◄ ◄ Scattered altocumulus cloudlets can indicate heavy rain or thunderstorms later

▲ A lowering cloud base with altostratus clouds indicates that a warm or occluded front is coming in around 8 hours

▲ Altostratus clouds before a warm front

mind the last known direction of the upper winds when using the crosswinds rule as the upper winds take much longer to change direction than the lower winds.

ALTOCUMULUS CLOUDS

Altocumulus clouds can indicate the approach of a weather front within 8 hours. Altocumulus clouds are light grey or white clouds which are made of little rolls or heaps. These small Altostratus clouds before a warm front elements are known as cloudlets and they come in a variety of forms, with gaps between each cloudlet where the air is sinking. Altocumulus clouds can also look quite messy, with fleecy patches.

Altocumulus clouds are often seen 4–8 hours before warm or occluded fronts arrive, where they are usually mixed in with layers of altostratus. These skies occur when fronts become older and less active, meaning that the rain is likely to be weaker and possibly intermittent. The wind may stay quite weak although windier conditions are possible. Again, use the crosswinds rule to double check whether a front is approaching as these clouds are often seen at other times

When cloudlets are arranged in waves or bands they often indicate the approach of a cold front, meaning that showers and thunderstorms are likely.

THUNDERY ALTOCUMULUS CLOUDS

Chaotic altocumulus cloud with turrets or bulging tops indicate thunderstorms and heavy showers within 12 hours. There are two special types of altocumulus that are important to be aware of. They are called altocumulus castellanus (photo page 39), which look like little turrets or towers (hence the

▲ Altocumulus floccus – 250,000 lightning strikes were recorded in the English Channel in the 12 hours after these altocumulus clouds were seen

name) and altocumulus floccus (opposite), which are little ragged scattered tufts with bulging tops. They usually mean thunderstorms in a few hours' time. They indicate that the air is humid and unstable. However, these thunderstorms usually cover large areas, unlike the local ones experienced from single cumulonimbus clouds.

The larger the turrets in the altocumulus clouds, the greater the instability and the more likely there will be thunderstorms later. The turrets are more obvious when the cloud is seen from the side, and thundery altocumulus clouds will sometimes have fallstreaks coming from the bottom of them too. They may be made of separated cloudlets, or their bases may be connected to each other. Just remember that any ragged or bulging altocumulus clouds indicate thunderstorms.

▼ Altocumulus castellanus

READING THE LOW CLOUDS

The low clouds are ideal for short term forecasting, but they are pretty much useless for forecasting long-term changes in weather. The low clouds are generally made of water in the temperate regions, although with sub-zero surface temperatures they may be made of ice.

▲ Cumulus cloud streets indicate more wind between the clouds than under them

◀ ◀ Thin cumulus clouds seen after mid-morning indicate fair weather for the rest of the day

▲ Cumulus clouds over the sea but not over the land indicate that it is windier out to sea

▲ Troughs indicate poor weather for around an hour and then the same weather as before

▲ Pannus clouds indicate rain in a few minutes

▲ Stratus clouds indicate that the weather will not change much soon

▲ Stratocumulus clouds also indicate that the weather will not change much soon

► Nimbostratus clouds usually indicate a depression

FAIR WEATHER CUMULUS CLOUDS

Thin cumulus clouds seen after mid-morning indicate fair weather for the rest of the day. Cumulus humilis clouds are thin cumulus clouds. They are always wider than they are tall. They can form early morning regardless of whether there will be good weather or not. It is only after mid-morning that these indicate that there will be good weather for the rest of the day, and there will almost certainly not be thunderstorms or showers later. These clouds will often form on sunny days.

▲ Fair weather cumulus clouds

Cumulus clouds are caused by rising air currents called updrafts. As air rises in these updrafts, it cools. When it is too cold the air cannot hold any more water, so the water condenses into clouds. Cumulus clouds have a flat bottom, as the height where the air condenses is usually very level. Also, cumulus clouds often have the bulging cotton wool shaped tops because the updraft is at its strongest in the middle, so it rises higher.

There is a daily cycle to fair weather cumulus clouds. In the early morning the sky is completely clear and there will be no clouds in sight. By mid-morning shallow cumulus clouds form that are a couple of hundred metres across. By early afternoon the clouds are at their largest. However, they are usually still wider than they are high. By mid-afternoon, the cumulus clouds start decreasing in size and will be much smaller and often fragmented. Just after sunset, there are usually no cumulus clouds in the sky at all.

These clouds will not often form near the coast. Instead, there will normally be an obvious line of clouds over the land indicating the sea breeze front which marks the start of these clouds. There will often be a sea breeze when these clouds form (see page 75). The cross-winds rule will not work if there is a sea breeze as the surface wind is going in a different direction to the way it would without a sea breeze.

The shade of cumulus clouds depends on the direction of the sun. If the sun is behind the clouds, they will look much darker than if the sun is shining on them. This is important as dark cumulus clouds do not generally mean that they are large or particularly moist clouds.

CUMULUS CLOUD STREETS

Cumulus cloud streets indicate more wind between the clouds than under them. Cloud streets are lines of regularly spaced cumulus clouds which can extend for hundreds of kilometres. They are made of horizontal convective rolls, on one side of which the air rises and, on the other, the air falls. These rolls fit together like cogs, so there are areas where the air is rising and cumulus clouds are formed, and areas where the air is falling which suppresses cloud formation. These cumulus cloud streets are generally a few kilometres apart.

▲ Cumulus cloud streets

Cloud streets are of interest because they cause wind bands, which are long strips of areas of stronger wind and lighter wind. Under the clouds there will be less wind, and in the clear sky areas there will be more wind and the wind will have veered a little. This is because the air that is rising has had longer to be slowed down by friction with the surface. The air that is falling has had less time to be slowed down by friction, so it is faster. The wind bands will move to the left as they are pushed by the gradient wind.

This is mainly useful for sailing as small differences in wind strength and direction could be the race-winning difference. It is, however, very useful for gliders as they can use the updrafts to stay in the air. Gliders can glide along cloud streets for hours – hence their name 'cloud streets'.

▼ Cumulus cloud streets

COASTAL CUMULUS CLOUDS

Cumulus clouds over the sea but not over the land indicate that it is windier out to sea. If it is clearer inland, but there are lots of cumulus clouds out to sea, this shows that there is an unstable air stream over the sea, indicating that the land is cooler than the sea. These conditions would mainly occur in autumn and winter.

It will be about two Beaufort forces windier outside the harbour and it will probably be very gusty. There will also often be showers out at sea but none over the land. Occasionally there may also be showers inland if there is a sea breeze, but this is unlikely as the sun is usually not warm enough for sea breezes to form in the autumn and winter.

▼ A view of cumulus clouds over a coastline from several miles inland

TROUGHS

Troughs indicate poor weather for around an hour and then the same weather as before. A trough is simply a valley in an air mass. Clouds pool up in the trough and poor weather is prevalent. A trough is an extension of a low pressure area, so has a lower pressure than the surrounding area. It is similar to a front but it does not separate two air masses. There will be showers and it will constantly be raining throughout the trough. There may also be thunder.

▼ A trough approaching

You can often be sure it is a trough by the obvious line where the shower clouds in the trough start. Before the trough, there are usually clearer skies, but in the trough, there are lots of shower clouds. The wind will also back a little. You shouldn't confuse it with a front as there won't be the same series of clouds in the lead up to the trough. It will also be gone completely in an hour without any clouds trailing behind it and there is no sudden wind veer.

Inside the trough it will be squally, showery and rainy. It will also be much cooler. Be careful as there is a small chance of tornados over the land and waterspouts over the sea.

STRATUS CLOUDS

Stratus clouds indicate that the weather will not change much. Stratus clouds are low, grey clouds that have few distinct features and they cover the whole sky. They are the typical 'dull day' clouds and are generally formed by maritime tropical air. They can build north of depressions (or south of them in the southern hemisphere) or in the middle of anticyclones during the winter in what is known as 'anticyclonic gloom'. They can also develop in the warm sector. They will often be seen covering the tops of hills or tall buildings. Stratus clouds often produce drizzle or rain if over hills or mountains.

They can be broken up into stratocumulus clouds by turbulence over rough land or by strong winds. Stratus clouds indicate that there will be no change in the wind or weather for at least a few hours, although sometimes they will thicken into nimbostratus clouds, which will lead to light rain or drizzle, especially on windward slopes.

▲ Low stratus clouds

STRATOCUMULUS CLOUDS

Stratocumulus clouds indicate that the weather will not change much for the time being. Stratocumulus clouds are a mix of stratus clouds and cumulus clouds. They have features of both stratiform and cumuliform clouds and they are a very common type of cloud, especially over the oceans. They are arranged in a single layer and can have gaps in them or can cover the whole sky. They can display features in the form of rolls, heaps or flat clouds with space between them. Stratocumulus clouds do not produce any precipitation and the winds will usually be light.

▲ Stratocumulus clouds during an anticyclone

Stratocumulus clouds are often present where there are temperature inversions like that of 'anticyclonic gloom'. This is because they can develop from cumulus clouds, which cannot extend above a temperature inversion, so they instead spread out sideways, covering the whole sky. As temperature inversions are stable, stratocumulus clouds indicate that there will be no immediate change in weather – and they can last for days. Smog can also be particularly prevalent in winter as the particles from industrial processes cannot escape into the atmosphere. Stratocumulus clouds are not typically found in depressions.

NIMBOSTRATUS CLOUDS

Nimbostratus clouds usually indicate a depression. Although nimbostratus clouds can cover all three clouds heights from the ground, they look like lower clouds, so are included in this category. Nimbostratus are grey clouds which can cover thousands of square kilometres. By definition there will always be precipitation with nimbostratus clouds as the prefix 'nimbo' means rain-bearing. Most winter rain is from this cloud.

Pannus clouds can be used to find out when this cloud will start raining (see page 53). If nimbostratus clouds are part of a front, then the rain will usually only last for about 4 hours. Nimbostratus clouds can also form if altostratus or stratus clouds thicken. To ascertain whether the nimbostratus clouds are part of a depression, use the last known direction of the upper winds when using the cross-winds rule.

▼ Nimbostratus clouds before a warm front

PANNUS CLOUDS

Pannus clouds indicate that rain will fall in a few minutes. Pannus clouds are fragmented or wispy grey clouds which form below the main cloud base. They form due to precipitation falling from above, which moistens the air until it can't hold any more water and clouds form. These clouds are also known as virga or fallstreaks when they are in their wispy form, and they look similar to the fallstreaks we see below cirrus clouds. Pannus clouds are fragmented due to the turbulent winds and will appear to move and change shape rapidly. They increase in number as the rain falls until they

▼ Pannus clouds under a cumulonimbus cloud – rain started 5 minutes later

cover the whole sky. The wind will also increase as more rain falls. Pannus clouds often form below nimbostratus clouds before a warm front but they also form below cumulonimbus clouds.

A rudimentary method of foretelling whether the pannus clouds have formed before a depression or not is to use the last known direction of the upper winds when using the crosswinds rule. Obviously, if you have been concentrating on the weather in the hours before the rain starts, you can be more confident in your prediction.

MAMMATUS CLOUDS (MAMMA)

Mamma indicate that a storm has just passed, or will arrive in minutes. Mamma, or mammatus clouds, look like lobes or udders that form under the anvils of cumulonimbus clouds; they are not technically a cloud in their own right, but a supplementary feature. They are caused by strong downdrafts, where pockets of cold moist air sink rapidly, against the main pattern of upward movement of warm humid air. However, the exact mechanism is still unknown. The air below the lobes is so turbulent that pilots are advised to keep well away from them.

Mamma come in all shapes and sizes, from spherical pouches to to tubes or just globules arranged in cellular formation. The lobes are usually made of ice and are normally about 500 metres tall and about 2km in diameter. The lobes last for about 10 minutes each, although a cluster of them can last for up to a few hours.

Mamma are generally seen below cumulonimbus clouds and indicate that there will be a shower very soon, with a good chance of thunder. They are mainly seen after a shower, which isn't particularly helpful for your forecasting, but they

do still look pretty spectacular. The longer and more dramatic the lobes are, the more violent the expected storm will be. Mamma are also occasionally found under other clouds, when they wouldn't indicate stormy weather, so don't always assume poor weather on seeing them.

▲ Mamma underneath a thunderstorm

Did you know?

Mammatus comes from the Latin word *mamma*, meaning udder or breast.

CLOUDS WITH VERTICAL GROWTH

5

Clouds which cover more than one cloud layer usually produce precipitation. Seeing clouds with vertical growth typically means that precipitation will arrive within an hour.

▼ Cumulonimbus clouds indicate heavy rain and possibly hail

◄ ◄ Towering cumulus clouds indicate showers within a couple of hours

▲ Thunderstorm clouds indicate strong wind, thunder and lightning, and heavy precipitation

TOWERING CUMULUS CLOUDS

Towering cumulus clouds indicate that showers will arrive within a few hours. Cumulus clouds that are taller than they are wide are called cumulus congestus clouds. Cumulus congestus clouds will have cauliflower-like florets on the top of them; if there is an anvil top or fibrous edges, then it is cumulonimbus and will produce heavier showers (see the next page). Shower clouds often form if the cumulus clouds are taller than they are wide before mid-morning. The earlier the cumulus clouds form, the heavier the showers are likely to be. These shower clouds usually only form in the summer but can be found at all times of year in tropical regions.

Shower clouds will grow when going up a slope. However, they will decay when going down a slope so if you see large shower clouds on the other side of the mountain, then when they reach you they are unlikely to be too ferocious. The

wind around showers will be strong as the shower cloud approaches but will tail off as the shower passes overhead. Often these clouds themselves do not produce rain showers, but they indicate the approach of cumulonimbus clouds in a couple of hours which do. If there are pannus clouds beneath the main cloud it is likely it will rain soon.

▲ A precipitating cumulus congestus cloud

CUMULONIMBUS CLOUDS

Cumulonimbus clouds indicate heavy rain and possibly hail. Cumulonimbus clouds are tall cumulus clouds which extend through all three cloud levels. They are the classic storm or shower clouds and will often produce thunder. Cumulonimbus clouds are bigger than cumulus clouds in width as well as height. They often have a base of around 20km across, which means that the showers can last around 30 minutes. Once a cumulonimbus cloud is overhead, it can be difficult to know whether it is cumulonimbus or nimbostratus – but the rainfall

▲ A cumulonimbus cloud with an anvil top

should be noticeably heavier if it is cumulonimbus.

Cumulonimbus clouds often have anvil-shaped tops. These anvil shapes occur because the rising water vapour in the cumulonimbus reaches the temperature inversion at the troposphere, and as it cannot rise any further, it spreads out, forming the anvil. Cumulonimbus clouds with an anvil are often decaying as the updraft that is creating the cloud is stopped in its progress by the temperature inversion. Cumulonimbus clouds will always have fibrous tops as the top section has turned to ice. The wind often reaches 30–40 knots when they pass overhead as the rain drags the wind downwards with it. Due to the rain forcing cold air down the temperature also decreases.

THUNDERSTORMS

Thunderstorm clouds indicate strong wind, thunder and lightning, and heavy precipitation. Thunderstorm clouds can be difficult to see when they are overhead but there will often be a clear edge where the cumulonimbus cloud starts. Usually there are sheets of rain visible, the cloud base is very turbulent and the clouds look quite ragged. A squall often occurs just before the clouds arrive, and the wind can reach 50 knots.

These thunderstorm clouds will not last long, probably a maximum of half an hour. However, other cumulonimbus clouds can merge, giving the impression that the cumulonimbus cloud is much larger than it is. These clouds are called multicell thunderstorms, where each cell is a cumulonimbus cloud. These multi-cell clouds can last hours, but they are quite rare. There are also larger mesoscale convective systems, which normally persist for several hours and can be over 100km wide.

Did you know?

The distance to the thunderstorm can be worked out by counting the time between the flash of lightning and the thunder. Every three seconds between the lightning and the thunder is one kilometre. Thunderstorms themselves travel at around 30 knots. Thunderstorms more than 15km away are not normally heard.

▲ Thunderstorm clouds – notice the sheets of rain on either side of the lake

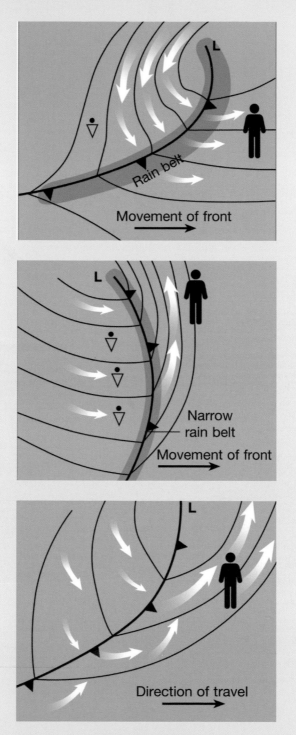

The cold front wind rule

WEATHER FORECASTING RULES

6

There are a few important rules of thumb to learn about weather forecasting. They won't always work but it is often worthwhile checking to see whether the cloud type confirms or disproves these predictions.

WIND BASED RULES

In the northern hemisphere, the lower wind always travels around a depression anticlockwise. Because of this, there is a rule called 'Buys Ballot's Law' which says, 'If you stand with your back to the wind, the centre of the depression is on your left.' This law was used for years by seafarers to avoid the worst of storms. It can be used to work out where the depression will pass. Take a bearing from your extended left arm and then repeat this again after a few hours. If there is clockwise rotation, the low will pass north of you, meaning you will probably stay dry. Anticlockwise rotation means it will pass to the south of you, so you are probably in for a soaking! Finally, if there is a steady bearing it means that the low is headed straight for you or is stationary. A barometer will tell you which of these it is.

Cold front rules

Weather fronts usually follow two scenarios; either it rains first and then the wind increases, or it happens the other way round. This causes the well-established rule:

> **Rain first, wind last**
> **Lash everything up fast.**
> **Wind first, then rain**
> **We can all relax again.**

This is because there are rain belts which are attached to cold fronts; either it rains ahead of the front or as it passes overhead.

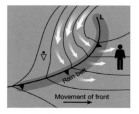

In the top scenario on the previous spread the rain comes first, then the wind. As the isobar spacing remains constant there won't be a change in the wind strength. However, once the rain belt passes the isobars get squeezed much closer together and the wind will be much stronger, possibly reaching gale force. If you have a barometer you would notice constant pressure until you reached the rain belt, where it would start to increase rapidly.

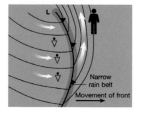

In the middle scenario on the previous spread the wind comes first, then the rain. The isobars start off closely spaced, so the wind force will increase rapidly. The rain will then start very close to the front, but as soon as it has passed the wind will drop down as the isobars are much more spaced out. The weather will then be fair, except for a few occasional showers. With a barometer, you would see that the pressure drops quickly at first but then remains constant once the front has passed. Notice also that the rain belt here is much narrower than the rain belt of the first scenario, so the rain wouldn't last as long but it is likely to be squally.

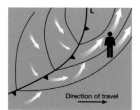

There is one more important scenario to be aware of with cold fronts. This is known as the calm front. You will notice in the bottom diagram that around the front the isobars fan out; this means that the wind will be much calmer around the front. Beware, this drop in wind is only temporary and the wind will increase again after the front has passed. Just remember that if the wind seems to drop off before, or as the rain belt in a front arrives, the wind will probably increase again to the same strength as before, so don't get complacent.

Using wind directions

The wind direction indicates the likely weather in Western Europe or the western United States. This is due to the air masses explained on page 10.

Wind direction	Expected weather
North-easterly to westerly	Cold temperatures and good visibility. In the winter hail, thunder and snow over western and northern coasts, but fewer showers in the east. In the summer, heavier showers occur in the east.
Southerly to westerly	Mild weather, with low cloud and drizzle or rain and poor visibility. Fog is possible on windward coasts or hills, although there is often fair weather in the lee of hills.
North-easterly to easterly	Extremely cold in the winter and hot in the summer. Rain is unlikely, although snow is possible in winter. Severe frosts are likely.
Southerly to easterly	Hot and dry, with poor visibility due to haze and pollution. This is the source of summer heatwaves.

VISIBILITY RULES

The different air masses all have their own humidity levels, as well as temperatures, so working out the type of air mass can help in forecasting the weather. Comparing the visibility is one effective way of identifying the air mass.

1. Improving visibility – after a period of fair weather with easterly winds, which always bring hazy weather, visibility will start improving steadily. This is a certain sign of deteriorating weather as it means that there is an air mass change coming. There is also likely to be a slow but steady fall in atmospheric pressure. This means that a low is approaching, which will bring a general change in weather.

2. Increasing haziness – if it becomes increasingly hazy during a calm spell in the afternoon then thunderstorms are likely later.

3. Poor visibility after rain – usually visibility improves greatly after rainfall as it washes all of the dirt out of the air. But if the visibility remains poor after rain then there are probably worse conditions to come. In the summer, showers are likely and possibly also thunderstorms.

4. Stars twinkling – if there seems to be an increasing twinkling of the stars then expcct the fair weather to end. This is because there is approaching cold air in the upper atmosphere, which is an extremely dry and clean air mass, causing the especially clear stars.

FOG

There are three main types of fog: land fog, sea fog and upslope fog. Each of these needs different conditions to form, so it is easy to differentiate between each one. Fog is dangerous for lots of outdoor activities, especially for people engaged in sports and in sailing; even driving is impacted by it, so it is important to know a little about fog.

Land fog

Land fog will clear when the ground is warmed by the sun or if the wind increases. Land fog, or more accurately radiation fog, is caused when the ground cools rapidly, which lowers the temperature of the air above it below the dew point. The water vapour then condenses into droplets. Fog is just a cloud that is touching the ground, although it is often formed by very different mechanisms. Radiation fog can be found in harbours, but not over the open sea. It is often found in valleys or over flat land. It is most likely to form when the air is moist and the sky is clear so that the heat can radiate away more easily. It is also more likely to develop when the nights are longer during winter, as this allows more time for the fog to form. The wind must be light otherwise turbulence from the wind mixes the air. An early dew indicates that the air is already humid so radiation fog is more likely.

▲ Radiation fog in a valley

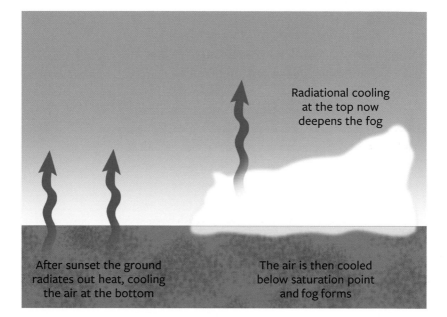

Radiational cooling at the top now deepens the fog

After sunset the ground radiates out heat, cooling the air at the bottom

The air is then cooled below saturation point and fog forms

▲ Formation of radiation fog

Radiation fog is often a very thin layer that is usually only a dozen metres thick. A higher place will often be out of the fog. The fog will thicken throughout the night, so at dawn the fog is usually at its thickest. It starts by forming a layer a couple of meters thick in river valleys and low lying fields, where there is more moisture and less wind to break up the fog.

Often the most useful thing about forecasting fog is working out the time when it will clear, as this generally governs when activities can commence. Radiation fog will disperse if there are winds faster than 5 knots. Also, the sun will 'burn off' the fog.

It does not actually burn off the fog; it warms the ground and thus the layer of fog above it also warms up, allowing the water droplets to evaporate. Rain will also clear fog. Radiation fog will almost always have cleared by mid-morning. Often the first sign that the fog will clear is that the sun can be seen clearly through the fog or that some blue sky is seen directly overhead.

Sea fog

Sea fog will clear when the wind direction changes or the water or land is heated up by the sun. Sea fog, or advection fog, is caused by wind carrying warm humid air over a colder sea. This air is usually maritime tropical air. The sea cools the air, which forces some of the moisture in the air to condense, causing fog to form. This fog can then be blown to other areas like beaches and harbours although it is unlikely the fog will penetrate more than a couple of kilometres inland. Sea breezes often blow advection fog inland, covering the beaches, but the fog is burnt off before it gets very far inland.

The first sign of advection fog is often light patches of mist over the coast before it spreads further out to sea. Also, if you notice your eyebrows getting damp and surfaces becoming wet then it often indicates that fog is on the brink of forming and just a degree of temperature change will trigger thick fog. If you see that smoke from ships' funnels or chimneys is crawling along the sea then fog is likely. However, if there are cumulus clouds over the sea then fog almost certainly will not happen.

Advection fog is the typical 'fog bank' type. It can be extremely thick and will often cover large areas of sea. It often clears along the coast due to air currents on coastal slopes and this is where the fog bank is especially noticeable. Argentia in Newfoundland has over 200 fog days per year as it is in the cool Labrador sea current but it has relatively warmer air. It is particularly prevalent in the spring and early summer as the air is much warmer than the sea.

Advection fog will also form more frequently inshore in winter and spring as the sea is coldest inshore, but in the summer and autumn it is coldest offshore, so fog is more likely out at sea. These temperature differences in the sea cause variations in the depth and thickness of the sea fog. Even a change in tide can alter the fog.

▲ An advection (sea) fog bank filling in from the sea

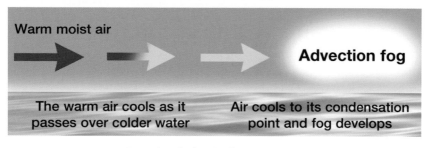

Warm moist air

Advection fog

The warm air cools as it passes over colder water

Air cools to its condensation point and fog develops

▲ Formation of advection fog

Warm fronts will also often bring fog as the air in the warm sector is warm and moist enough for fog to form. This fog will build in the hours before the front passes, whilst it is raining, as the warm and moist air is brought down onto the water. Then, when the cold front arrives, the fog will slowly start to disperse.

Advection fog does not disappear over the day like radiation fog and it can even persist for several days. It is especially difficult to predict when advection fog will clear.

A wind shift will often clear advection fog as the new wind is unlikely to have the same humidity and temperature. An increase in wind strength may also help the fog dissipate, although you will often have to wait until the next cold front for the fog to disappear. Also, if you are at sea and are moving into warmer waters, the fog may clear.

As with land fog, often the first sign that advection fog will clear is that the sun can be seen easily through the fog or that some blue sky is seen directly overhead. Advection fog will often clear downwind of islands as the heat from the land warms it up enough for the fog to evaporate.

Upslope fog

Upslope fog will clear if the wind direction changes or you move to the leeward (downwind) side of a slope. Upslope fog is, quite obviously, fog that is formed up a slope. Sloping terrain lifts moist air, which cools until it reaches the dew point. It looks like fog from inside but looks like stratus cloud from above or below. This fog, unlike radiation fog, can sustain itself in higher wind speeds although winds of over 12 knots often cause it to disperse into stratus clouds instead. However, a slow and steady wind can often feed the fog until it becomes thicker and longer lasting.

▼ Formation of upslope fog

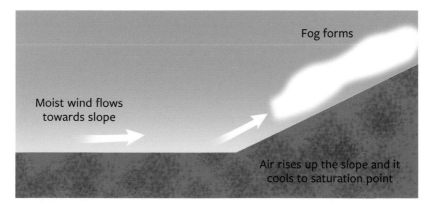

Fog forms

Moist wind flows towards slope

Air rises up the slope and it cools to saturation point

▲ Upslope fog on the windward side of a mountain

Upslope fog normally begins to form high up in the mountain and then thickens until it almost reaches the valley floor. Upslope fog often occurs on sunny days as the mountain air heats up faster than the air from the foothills, causing lower pressure over the mountains. The air will then flow to the lower pressure, rising up the mountain slopes. This process often causes a sea of clouds in the valleys below.

Be careful when the humidity is high and you are climbing up the windward side of a mountain as upslope fog could form at any time. Upslope fog will disperse if the air mass changes, causing the temperature or humidity of the air to change or if the wind direction changes. There will be no upslope fog on the leeward side of mountains, so it is important to choose a route carefully. It is particularly common at Mount Emei in China, where there is fog for 300 days per year, and also on the eastern side of the Rockies in the US.

LOCAL CONDITIONS 7

Sometimes there can be wind at your location, but none 5km away or rain on one side of a hill but not the other. There are many factors which affect local weather conditions.

SEA BREEZES

Sea breezes are of paramount importance to the coastal sailor but can also be the difference between a warm day and a cold windy day on the beach. Sea breezes are winds which blow from the sea to the land. They only occur on sunny days and there is a clear process to how they start:

1. The sun warms up the land which in turn warms up the air.
2. This air expands as it warms and rises (a couple of degrees temperature difference is enough to start the process).
3. This causes higher pressure a few hundred metres above the land.
4. Air will then flow out to sea as air always moves from areas of high pressure to areas of low pressure.
5. The air out to sea moves down to take the place of the air that is moving onshore.
6. This air that is moving onshore is the sea breeze.

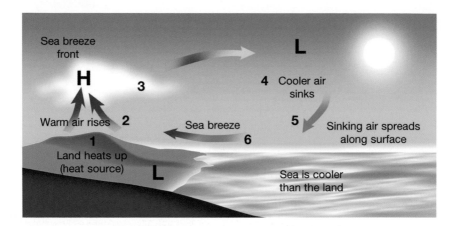

Sea breeze front

H

3

4 Cooler air sinks

L

Warm air rises **2**

Sea breeze **6**

5 Sinking air spreads along surface

1

Land heats up (heat source)

L

Sea is cooler than the land

▲ Circulation of air in a sea breeze

Once the sea breeze starts it continues to grow quickly, spreading several miles out to sea and far inland. The further inshore it reaches, the stronger the wind on the coast will be. In early summer the sea breeze can reach up to 50km inland and a similar distance out to sea, and it can reach up to 25 knots on the coast. The wind strength generally increases until mid-afternoon and it will stay relatively constant until sunset, when the sea breeze quickly dies away. Throughout the day the sea breeze will veer about 60 degrees.

Where the air is rising there will be a distinct line of cumulus cloud, called the sea breeze front, which is the main sign of a sea breeze. The taller the cumulus clouds are, the stronger the sea breeze. The sea breeze front is actually a cold front as fronts are always named after the air behind them – colder maritime air in this case, which is retreating inland at a few miles per hour. The wind below the sea breeze front is almost completely calm, which is very unlike an ordinary cold front. However, temperature inversions act as a lid to sea breezes, which significantly reduce their strength.

▲ A sea breeze front

Sea breezes on islands and peninsulas

Sea breezes can also occur on islands and peninsulas. The sea breeze starts just like it would on the mainland and flows inshore, cooling the island or peninsula. The sea breeze then meets itself in the middle and a cumulus cloud will form there. As islands or peninsulas are often small, there is usually not enough heat to maintain the sea breeze, so it dies. Then after a few minutes, the sun warms up the island or peninsula enough again so that the sea breeze continues. This pulsating sea breeze will continue throughout the day. Islands and peninsulas that have pulsating sea breezes like this are usually around 10–50km across. The larger the island, the longer the wind will continue to blow. Larger islands and peninsulas often have showers in the afternoon due to the amalgamation of two sea breezes causing larger cumulus clouds than normal.

An example of this is the Isle of Wight in the UK. Early in the morning, the sea breeze will pulsate as the island warms up and cools down. This situation is made more difficult as the island is only a couple of kilometres from the mainland. The sea breeze on the island will compete with the sea breeze on the mainland leaving an area of calm between the two. As the day progresses the sea breeze on the mainland will be stronger so the sea breeze continues just like it would have without the island. It is similar with most of the Caribbean islands, such as Antigua.

Cornwall in the UK is a good example of a peninsula which has a pulsating sea breeze. The sea breezes start as they would normally. However, as they flow inland they meet in the middle, causing a line of cumulus clouds. When the sea breezes meet in the middle the sea breeze will die for about 15 minutes and then the land will warm up enough for the whole process to start again. It is similar with the Cherbourg peninsula in France and the Baja peninsula in Mexico.

Interestingly, sea breezes can also form on lakes and are similar to breezes on islands as they will pulsate throughout the day. A lake with a 5km diameter will normally only be able to maintain a sea breeze for about half an hour before it runs out of warm air.

When the gradient wind is offshore

Forecasting sea breezes is relatively straightforward when there is no gradient wind (lower wind without the sea breeze effect). But when the gradient wind is factored in, things are different. Contrary to popular belief, an offshore wind helps a sea breeze form even though it is going in the opposite direction to the sea breeze on the surface. This is because the gradient wind helps the return flow of air aloft. As can be seen in the figure on page 76, the sea breeze is a circular flow of air, not a one way flow, and if the gradient wind is pushing the return flow of air it aids the circulation of the sea breeze. The sea breeze we feel can flow right underneath the gradient wind unless the gradient wind is faster than about 25 knots. If the gradient wind is blowing parallel to the shore, then there will still be a sea breeze, but the sea breeze will be blowing almost parallel to the coast.

Whenever the wind is offshore there will be a zone of total calm where the gradient wind is fighting the sea breeze. Sailors need to be careful of this band of calm, which starts right next to the coast and will move further out to sea as the day goes on. Crossing it is a risk that sailors often have to take as the sea breeze will be stronger nearer the shore.

When the gradient wind is onshore

When the gradient wind is onshore, if it is faster than about 5 knots, then a 'real' sea breeze will not form, although there will still be an onshore breeze which could be confused with a sea breeze. Wind flows from areas of high pressure to areas of low pressure, so in an onshore wind the pressure will be lower over the land. When the land heats up, it will lower the pressure further so this will cause the wind strength to increase by up to 5 knots. This is one of the reasons why the wind will increase near the coast on a hot day. This is called thermally-enhanced wind. Due to the lower pressure over the land the isobars will shift slightly. This causes the wind to veer a little.

Although it may be hard to tell the difference between a sea breeze and a thermally-enhanced wind there are some key differences:

1. In a thermally-enhanced wind, the wind does not die down as it sets in, unlike the sea breeze, and there is no band of calm that moves offshore throughout the day.
2. In a thermally-enhanced wind, the wind veer is usually much less than that of a sea breeze.

> 3. In a thermally-enhanced wind, the increase in wind speed covers an area a few kilometres wide and is the same strength throughout, unlike the sea breeze, where the wind is stronger nearer the shore.

There are often cases, especially in late summer, when the sea breeze is fighting against an onshore gradient wind, which can cause very light and shifty wind conditions. Often the sea breeze will win for about half an hour, only to be taken over by the gradient wind again. In these circumstances, it can be difficult to tell whether you are in the sea breeze or not. However, cumulus clouds over the shore will confirm whether or not you are.

▼ Thermally enhanced wind

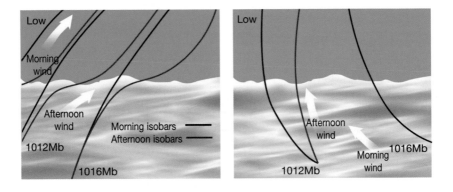

Remember

Overall, if you are not a sailor it does not make too much difference whether it is a sea breeze or not. Whilst this extra knowledge is interesting, it is never included in forecasts. However, it could help you to choose the most sheltered beach when you are on holiday and is also critical for pilots. Below are the most important points for non-sailors to note:

1. Sea breezes occur on sunny days when the land warms up quicker than the sea; sea breezes are a convection current.
2. The sea breeze front marks the furthest extent of the sea breeze inshore.
3. Sea breezes strengthen and veer as the day goes on.
4. Normal sea breezes are strengthened by an offshore wind and prevented by an onshore wind.
5. Sea breezes can reach up to 25 knots in the right conditions.
6. A very strong offshore wind can stop a sea breeze.

Of course, if you are a sailor this information is essential, yet many top sailors seem to ignore the details of a sea breeze and assume it only comes in the summer if there is an onshore breeze and it is sunny.

The land breeze

At night there is often a reversal of the sea breeze. This is known as a land breeze. A land breeze is the opposite of a sea breeze, where the land cools rapidly and the air becomes cooler than that of the sea. It then flows out to sea. These land breezes are much weaker than sea breezes; they generally only reach about 3 knots and they extend no more than about 2km out to sea and about 5km inland.

Land breezes are also extremely thin, generally no deeper than about 10 metres. However, if a katabatic wind (see page 87) is added into the mix these land breezes become a little stronger – up to about 10 knots in the UK. In mountainous areas, such as some places around the Mediterranean, katabatic winds can reach up to 50 knots in the valleys. The amalgamation of the land breeze and katabatic wind is known as a nocturnal wind. This will be funnelled down valleys or estuaries, where they reach their strongest. However, when this happens, the land breeze in nearby areas will be starved of wind. It is best to think of the nocturnal wind as a 'drainage' wind, where the cold air flows down hillsides like water would.

GUSTS AND LULLS

Gusts are areas of stronger wind and lulls are areas of lighter wind. Gusts look like darker patches over the water and lulls look like lighter patches.

Gusts and lulls around clouds

These gusts are harder to see as they cover large areas. However, over a period of a few minutes there is a noticeable difference in wind speed and water shade. Gusts and lulls around clouds are caused by thermal overturning, where areas of air are heated, which then rise and are replaced by colder air from higher in the sky. Just like cumulus cloud streets, below cumulus clouds there is lighter wind that is more backed. This is because the air here is rising and has been in contact with the water for much longer than the air in other areas, so it has been slowed down more by friction. In the areas where there are no clouds the air is falling. This means that the air has been in much less contact with the sea, so it has been slowed down less by friction and is less backed. These gusts will usually arrive at regular intervals, but it is very easy to predict when they will arrive as the lulls are present under the clouds and the gusts are under the clear sky.

The larger the cumulus clouds, the larger the wind shifts, gusts and the intervals between the shifts. When sailing, try to avoid these clouds and head for the clearer areas. This only applies to cumulus clouds which are not producing any precipitation; precipitating clouds often have their own wind.

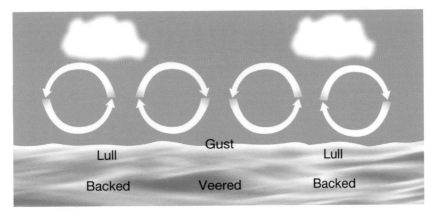

▲ Gusts and lulls around cumulus clouds

Gusts due to mechanical mixing

'Mechanical mixing' is where the wind mixes at different layers as the surface of the sea or land is not perfectly smooth. Gusts and lulls formed by mechanical mixing will be random and there is no rule to the pattern of wind shifts during the gusts. These are the most easily seen gusts and are especially prominent on lakes and in offshore winds at sea. It is almost impossible to see which feature upwind caused the gusts and lulls. However, rougher areas like forests or hills cause many more gusts and lulls. Gusts can also be formed down valleys and lulls behind hills.

THE WIND AND THE COAST

Offshore winds

When the wind is onshore the sea will be much rougher, but the wind will be stable and not gusty. However, when the wind is blowing offshore, the sea conditions will be slight as there is not much fetch (the length of water over which a wind has been blowing). It is important to know the impact that the

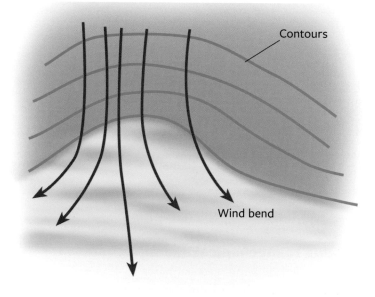

Contours

Wind bend

▲ A wind bend in offshore conditions

land has on the wind. If there are valleys leading down to the coast, then expect more consistent conditions and gusts up to 50% stronger downwind of the valleys. This is because the wind accelerates down valleys towards the sea. These gusts are in lines extending down from the valleys and are called wind bands. The colder the air, the stronger this effect, as cold air is denser.

When the wind is coming off higher ground expect shifty conditions due to turbulence in the wind. In some places under cliffs the wind may be going in the opposite direction to the rest of the wind. This is called an eddy. The wind will also spread out of the valleys behind the higher ground. This will cause a wind bend and if you sail into one of these and then tack in the middle, you will get a lift on each tack, which means you can sail further upwind. Because of the wind shifts and gusts be careful when going inshore to avoid waves, as the wind conditions may be difficult. This, of course, doesn't apply to a power boat, where the best practice would be to travel as far inshore as safely possible. The wind bands caused by this can be seen very easily and look like much darker patches on the water. If you are flying kites in an offshore wind on a beach, go downwind of a valley for stronger and more constant winds.

▼ Wind bands downwind of valleys in an offshore breeze

Winds parallel to the shore

When looking upwind and with the wind almost parallel to the shore, if the land is on the left, then the wind will be 25% stronger. However, if the land is on the right the wind will be 25% weaker. This is because wind going over the land is backed much more than the wind going over the sea due to the extra friction of the land. Looking upwind, when the land is on the left the wind is more backed so it converges with the wind over the sea, which is less backed. This makes the wind stronger in this zone of convergence. The convergence zone normally extends to about 2–5km offshore.

The opposite occurs when the land is on the right when looking upwind. As the wind over the land is more backed it diverges from the wind over the sea, causing areas of lighter wind offshore. Normally during the afternoon in the warmer months of the year, this divergence effect is countered by a thermally-enhanced wind. This effect is of no significance to non-sailors, but is crucial for sailors and cruisers to know about. So remember, if there is land on the left when looking upwind, go left for more wind.

Topography also affects the wind strength and direction enough that even coastal walkers will notice. Often in gales the strongest wind is not far out to sea or up a tall mountain but next to a steep headland. When the wind is going around a headland it often compresses the wind rather than shelters it. The compression of the air flow causes stronger wind just off the headland. The higher and more pronounced the headland the greater the effect of compression. Even low-lying headlands will cause noticeable increases in wind strength.

Next time you are walking on the coast, if the wind is parallel to the shore look out for darker areas of sea or increased white horses around the headland. This is useful for sailors, as hugging the shore could result in increased wind strength compared to far out to sea, reducing passage times. The wind will also increase in strength when it goes over a hill, and at the top of a hill the wind can be almost twice as strong. This is why picnics on the tops of hills aren't always a good idea!

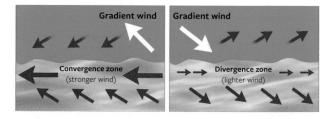

▶ Convergence and divergence over the coast

KATABATIC AND ANABATIC WINDS

Katabatic and anabatic winds occur around mountains and are especially important for hill walkers and mountain climbers to understand. Katabatic means 'going down' (from the Greek 'kata' = down). Katabatic winds occur at night and, like the land breeze, can be thought of as drainage of cold air flowing down the valleys just like water. The mountains cool down faster than the sheltered valleys, cooling the air with them. This air is colder and denser than the air around it, so it slides down the mountains and cools the areas in the bottom of the valleys. The steeper the slope, the stronger the wind will be.

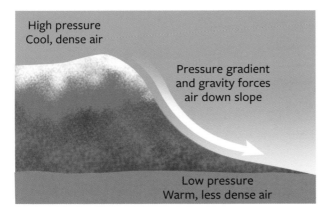

High pressure
Cool, dense air

Pressure gradient
and gravity forces
air down slope

◄ Formation of a
katabatic wind

Low pressure
Warm, less dense air

Anabatic winds are, by contrast, much rarer upward moving winds (from the Greek 'ana' = up). During the day the air in the valley is heated by the sun and then flows up the slope, often causing cumulus clouds at the top of the mountain, obviously marking the extent of anabatic wind. Upslope fog may also form here if the air is moist.

Katabatic winds can become very strong if cold dense air accumulates over high land such as a plateau and then flows downhill. These winds are extremely cold and can be very strong. Katabatic winds are extremely common in Antarctica and have reached almost 200mph – no wonder it is called the home of the blizzard. In the Rhône valley the Mistral is an example of a katabatic wind like this. However, in the UK katabatic winds never get very strong at all; they probably don't reach more than about 10 knots. Often people think that they are experiencing a katabatic wind when they are actually experiencing either the downdrafts from a cumulonimbus cloud, or the chaotic turbulence beneath a cliff in an offshore breeze, which often produces strong gusts from random directions.

The lower wind will often flow through the valleys resulting in very different wind directions to what was forecast. Knowing how these mountain winds behave is essential for all hill climbers, walkers and the like.

RELIEF RAIN

Relief rain (orographic precipitation) is precipitation that is produced when moist air is forced to rise because it encounters a mountain. As the air rises it cools and the water vapour condenses into clouds. These clouds then produce precipitation that falls on the upwind side of the hill. The air, having lost most of its moisture, then flows down the other side of the mountain.

As this air is much drier than the air on the other side of the mountain, less precipitation falls on the downwind side. This is called a rain shadow. Mountains do not have to be very high to cause a significant rain shadow; the Cairngorms in Scotland, which are only 1,000m high, experience about half the rainfall of the western Scottish mountains of the same height. More extreme examples include Death Valley and the Gobi Desert.

There is also another effect on the downwind side of mountains which is referred to as the Föhn effect. This effect causes warm dry winds that can become quite strong. In the Rockies, they are known as Chinook winds. They are caused because saturated air cools slower than dry air. The saturated air then rises up the mountain, cools slowly and sheds its moisture. When the drier air flows down the other side of the mountain it warms more quickly, causing a temperature difference of about 1 degree Celsius every 300m of mountain height. This air is also well known in the Alps where places downwind of the mountains experience unseasonably warm temperatures.

◀ Relief rain and Föhn winds

THE WIND AT NIGHT

Thin temperature inversions form overnight as the land cools quicker than the air above it. These temperature inversions cause the air to become very stable. This stability means that there is not a continual transfer of wind downwards. Normally air is unstable, meaning the warmer air is below the colder air. This air rises allowing new wind to replace it. When the air is stable, the wind cannot easily penetrate this thin layer of stability, so although the wind may be blowing high up in the sky, there will be no wind on the surface. This is the cause of still mornings when the wind then picks up later in the day.

The temperature inversion is heated by the sun, so the wind picks up early in the morning as the air becomes unstable. By mid-afternoon, the wind will have started to drop again as the ground is radiating away more heat than it is receiving from the sun. After sunset, the surface wind often drops to zero again. An indicator of a temperature inversion is that air pollution is often trapped below the inversion, leading to poor visibility.

However, temperature inversions like this never form

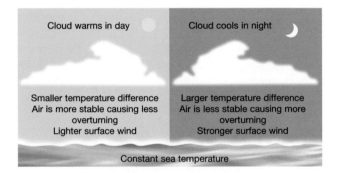

▶ Differences between day and night

over water, as the water takes much longer to cool down. Instead you actually get the opposite, with the wind strength increasing during a cloudy night. This is because the cloud tops cool down rapidly in the night, meaning that it is much colder far above the sea, than in the air just above it. This temperature difference causes the air to become very unstable as the cold air wants to sink. The result is more overturning of the air, whilst the wind increases as it receives more of the wind from up high.

FORECASTING SNOW

Snow is difficult to forecast because it requires knowledge not only of whether there will be precipitation, but also the temperature on the ground and at different altitudes. Snow can form at one altitude and then move into a warmer layer and turn to rain and can then even turn back into snow again, all before it reaches the ground. This makes the forecasting models used by weather forecasters very difficult to use. Also, it can often be raining in one area and just a few miles down the road there are a good few inches of snow settling in the fields.

Although water normally freezes at 0 degrees Celsius, the water droplets in clouds will not be able to freeze until the temperature is below about −10 degrees Celsius, as they have no particles to freeze around. This means that sometimes we might get temperatures of below freezing on the ground, yet rain is falling instead of snow as the clouds are just too warm. There is no easy way for us to work out the cloud temperature ourselves, so don't be surprised if the conditions seem to be perfect for snow, yet all you get is freezing rain.

If you have a thermometer handy you can work out whether any snow falling from above will reach the ground in a frozen form. Contrary to popular belief, the temperature doesn't need to be below freezing for snow to reach the ground; in fact snow can fall in up to 2 degrees Celsius. If the temperature is 3 degrees then you are likely to get sleet or partly melted snow, depending on the temperature above you. It also can never be too cold to snow, unlike popular belief. This is often thought to be true because the coldest conditions occur with anticyclones, when conditions are stable and precipitation is unlikely.

Assuming the temperatures are right for snow, you forecast it using the clouds in exactly the same way as you would forecast rain, although cumulonimbus clouds are rarer in the winter, so you are more likely to get snow from nimbostratus in depressions.

When the icy wind warms, expect snow storms

This fairly well known piece of weather folklore is actually quite accurate. This is because during times of cold weather there will often be a series of high and low pressure systems. The temperature will be particularly low in the high pressure systems, due to the heat escaping from the clear skies. However, when the weather starts to warm it means that a low pressure system is coming, bringing precipitation, so assuming the temperature is still low enough, heavy and prolonged snow can be expected.

In the Scottish mountains almost any front or depression will bring snow in the winter, but in the lowlands of Britain a warm or occluded front will be the least needed for snow to form, although a depression passing just to your south could bring snow too (use Buys Ballot's Law, see page 65, as well as watching out for the cirrus and other clouds that indicate a depression). Cold fronts, despite their name, don't often bring snow, though occasionally the rain will turn to snow behind the front.

If it is sleeting outside there are a few factors which will impact on whether it turns into pure snow. Most important is whether night is approaching, as that will cool things down sufficiently, but falling pressure, a colder wind or even an increase in the intensity of the sleet could result in snow.

The snow will eventually melt because either the ground is too warm or the air is too warm. Snow will last a long time if it makes it past the warm sector of a depression and through the cold front. This is because there are likely to be a few days of freezing clear days before another warm front arrives. This warm front may temporarily bring more snow, but then the warm air will arrive and the snow will start to melt rapidly.

FORECASTING TEMPERATURE

Forecasting temperature is difficult for amateur forecasters who don't have time for calculations involving sun intensity and other complicated factors. However, using a little information it is possible to hazard a guess at what the temperature will be like soon – although generally it will only be as precise as whether it will be warmer or colder tomorrow.

Stratus type cloudy conditions generally mean that the temperature will be cooler in the day as they reflect the heat from the sun, which would otherwise have reached the surface. However, at night the opposite is true. The heat from the earth is radiated back out into space, but any cloud cover will reduce this effect, causing a warmer night. The same is true on cloudy winter days, which will be warmer than clear winter days. Generally depressions, with their cloud cover, have milder conditions, and anticyclones have more extreme temperatures. However, it is important to remember that anticyclonic conditions often cause extensive stratocumulus clouds, especially in the winter, and there are also much milder conditions under these.

The wind direction is also quite indicative of what the temperature will be. If you know where the wind is likely to be shifting, you should be able to work out an approximate temperature for tomorrow (you might be expecting the wind to veer west as a depression is approaching or you may be noticing a general tendency of a wind shift). Use the table on page 67 to work out what temperatures to expect with which wind directions.

▲ A calm evening with a temperature inversion

You can also work out what the weather will be like at different elevations if you know whether there is a temperature inversion or not. Some of the telltale signs of a temperature inversion are layered stratus, stratocumulus clouds or fog, smoke not rising, windless nights (especially if the day was breezy) – if you see a couple of these signs then you can be confident that there is an inversion.

If there is a temperature inversion it means that it is getting warmer as you ascend, rather than cooler. This is useful for mountaineers and hill walkers as they will know that it is unlikely to be as cold as expected at higher altitudes. These are the conditions where you get the 'sea' of clouds, with the hills or mountains rising up like islands and blissful sunshine above the clouds; this is known as a cloud inversion.

USING A BAROMETER

A barometer is an instrument that measures air pressure. Air pressure is simply the weight of the air above you and it is measured in millibars. Barometers are indispensable aids for anyone who wants to forecast the weather. However, the barometer only tells you what the pressure is at that instant. This has nothing to do with the current weather and it is perfectly possible to have thunderstorms in extremely high pressure and blissful sunshine in extremely low pressure. What is important about the barometer is the rate of change in air pressure. The prophetic text stating 'Fair', 'Change' etc on the barometer can usually be ignored.

A great deal about the weather can be worked out from the rate of change of the barometer:

A slow and continuous fall in pressure means poor weather will persist. The end of a long period of fair weather is indicated by a steady reduction of pressure. It means that either a series of minor low pressure systems or an extensive major low is approaching.

Short term pressure changes indicate unsettled weather. There is likely to be foul weather consisting of strong wind, rain or fog, but it is unlikely that each of these will last more than half a day if the pressure keeps rising and falling. This is because there is a series of lows passing in quick succession but between each of the lows is sunny weather, caused by what is called an intermediate high. An impending change can be recognised by an overall rise in pressure, indicating settled weather soon.

Slowly and steadily rising pressure brings lasting weather improvements. This foretells that a high pressure system is approaching and the slow pressure rise points towards the high being very large, indicating a long period of fine weather but a lack of wind.

Sharply rising pressure brings strong wind. Just because it is sunny it doesn't mean that the weather will stay fair. These high pressure gales can be particularly prevalent in the summer and they can last for several days, whilst bringing blue sky. A rise of over 1 millibar per hour is a sure sign of a high pressure gale.

As wind is a direct product of a pressure gradient, a quick change in the pressure indicates strong wind as the pressure gradient is steep. A pressure change of about 5 millibars in 3 hours will result in a strong breeze with winds of around 25 knots. A change of 8 millibars in 3 hours will result in a gale and a change of 10 millibars in 3 hours will result in a storm. It helps if you write down the pressure every hour as that will show you the trends more easily and you won't forget what the pressure was earlier. Although failing that, you can twist one of the dials to align with the current reading, which will show you the pressure change at a glance when you check it on the next hour.

FORECASTING WAVE HEIGHT

Though not as useful for people far away from the coast, it is very useful for sailors, surfers and other beachgoers.

There are two types of waves: swell waves, which are made somewhere else and have moved in, and wind waves, which are produced by winds locally. Wavelength (the distance between wave crests) and wave height depend upon the wind strength, current, how long it has been windy for, the water depth and the fetch (the distance upwind to an obstruction like the shore).

Swell waves are the waves that surfers usually surf. They are the large waves that are often present on an otherwise completely calm day. They are generated in the middle of oceans over a period of days and they last for about as long as they have taken to be produced. They can move thousands of kilometres across the oceans and they often have wavelengths hundreds of metres long.

Swell waves are extremely useful for forecasting an advancing depression. Swell waves often move faster than weather systems, so large swell waves often indicate that a depression is coming. They are mainly useful to tell us that there is not a depression on its way. If there are all the signs of an approaching depression, but there is not any swell ahead of it, then it is likely that your forecasting is incorrect. Also, threatening skies without any swell often means there is just a thunderstorm rather than a depression. Remember, if there is no swell ahead of threatening skies it suggests that the bad weather will be temporary. If there is swell, then a depression is probably approaching.

Wind waves are locally produced waves and they can be created within an hour. Usually, the longer the fetch, the taller the wave height. If the wind has been blowing for 2 hours at 10 knots, the wave height will be approximately 25cm so long as the fetch is more than 8km. If that wind continued for 6 hours more, then the wave height would be 50cm assuming the fetch is more than 40km. See the chart below for more wave height examples.

Wave height is also impacted by water currents. If the current is flowing in the same direction as the waves, the waves will be flatter, but if the current is travelling in the opposite direction to the waves, the waves will become steeper. This is because in wind-against-current situations the waves are slowed, and as waves cannot easily lose energy, the speed energy of the waves is transformed into making taller and steeper waves. In tidal waters, this effect is known as wind over tide.

Wind Speed (knots)	Time blowing (hours)	Height (m)	Minimum fetch (km)
10	2	0.25	8
10	6	0.5	40
20	5	1	40
35	5	2	50
35	20	6	430

When waves reach shallow waters they start to behave differently. They slow down when the wave starts 'feeling' the ground, thus increasing wave height and decreasing wavelength, which makes the wave steeper. When the water is too shallow the wave eventually becomes too steep to be able to stay upright, so the crest falls down. This is what occurs when waves break.

Waves will start feeling the ground in depths less than half a wavelength, which is normally near the shore. However, issues arise when the wavelength is hundreds of metres long, as it so often is in the North Atlantic. The Bay of Biscay is a perfect example of when waves with 500-metre wavelengths start to feel the bottom of the relatively shallow (200m deep) Bay of Biscay. This causes the waves to become much taller, and combined with an Atlantic storm the fearsome Biscay conditions occur.

Waves travel in straight lines until they reach an obstruction, like a breakwater. The part of the wave that doesn't hit the breakwater will, instead of continuing in a straight line, diffract (bend) around the breakwater, meaning that even if you are behind the breakwater, the waves will still reach you. This is also obvious in bays with a narrow entrance, such as Lulworth Cove in the UK. Waves will also bend around obstructions like islands, causing a confused sea state.

▲ Diffraction around a breakwater

WEATHER LORE

8

Weather lore or weather proverbs were considered the most accurate way of forecasting the weather for centuries. Some of them are still well-known phrases and many of them have some truth in them.

> **Red sky at night, sailor's delight**
> **Red sky in the morning, sailor's warning**

This is the most well-known of the weather proverbs. It appears in the Bible, so was known at least by the 1st century AD. It works due to the position of the sun and the clouds. For a red sky to occur, it must not be cloudy on the side of the sky where the sun is. However, there have to be clouds in the other side of the sky onto which the sun can shine its red light. If there is a red sky in the evening it means that there must be clear weather to the west, but poor weather to the east. This means that the weather is improving as weather systems generally come from the west. In the morning the opposite occurs: the poor weather is to the west and the good weather is to the east; the red sky indicates deterioration in weather.

This proverb is only correct in the mid-latitudes where the weather systems generally travel west to east. Though it can be accurate, it is possible that when the sun is setting the sky can clear briefly in the west, allowing the sun to shine on the other clouds, and yet the poor weather is still to come. Similarly if the sun is rising and there happen to be a few clouds in the west, then it will look like poor weather is coming, when in fact there may be fair weather.

Mares' tails and mackerel skies make tall ships carry low sails

Here 'mares' tails' refers to cirrus clouds and 'mackerel skies' refers to altocumulus or cirrocumulus clouds. This sky often means that a depression is coming. These clouds are the first indication of a warm front, so there will be no rain for the first 12 hours at least. As depressions are windy, it indicates to ships that they should lower their sails in time for the storm. This is quite accurate as there are almost always cirrus clouds and sometimes cirrocumulus or altocumulus clouds before a depression. However, these clouds can sometimes be seen without a frontal system approaching.

A halo around the sun or moon means rain or snow coming soon

The refraction of light in the ice crystals in cirrostratus clouds causes halos. Cirrostratus clouds are often present before a depression approaches and as we know, depressions bring rain or snow. Cirrostratus clouds are not often seen unless a depression is approaching, so this weather proverb is usually quite accurate.

Long foretold, long last
Short notice, soon past

If a weather system is first seen a long time before it actually arrives, it indicates that it is travelling slowly, so it will pass slowly. However, if a weather system is only first seen just before it arrives, then it is travelling fast, so it will pass quickly. The same applies to showers, which are only seen for a few minutes before they start; they will last for only a few minutes. This weather proverb is usually correct and is one that is extremely useful to know.

The sharper the blast, the sooner 'tis past

This one basically means that the more ferocious the poor weather and the quicker it sets in, the shorter it will last. Strong thunderstorms usually last less than an hour; however, in a warm front or even in the warm sector the rain will last hours or days but will be much lighter.

A sunny shower won't last an hour

A single cumulonimbus cloud (shower cloud) is never larger than about 15km. Often cumulonimbus clouds join to form a multicell, but multicell clouds do not have sunny gaps between them. This means that if there are sunny breaks between showers, the showers should pass over in much less than an hour. This is very accurate correct due to the nature of cumulonimbus clouds.

When clouds appear like rocks and towers, the earth's refreshed with frequent showers

This proverb is fairly obvious. It is referring to the towering cumulus clouds, cumulus congestus and cumulonimbus clouds. Cumulonimbus clouds look like large rocks in the sky and they tower up kilometres high. If you see cumulonimbus clouds upwind of you, try to find shelter soon as they usually bring pretty ferocious showers and even thunder and lightning.

When the new moon holds the old one in her lap, expect fair weather

Just as the moon reflects light onto the earth, the earth reflects light onto the moon. This is called earthshine and allows us to see the dark part of the moon occasionally. When the dark part of the moon can be seen next to the new moon, it indicates that the air is very dry. As weather fronts are damp, if you cannot see the dark part of the moon a depression may be coming. Also, showers require a humid atmosphere, so if the dark part of the moon can be seen then showers are unlikely. However, a trough could catch you out by surprise.

If the goose honks high, fair weather
If the goose honks low, foul weather

This proverb is referencing how high the geese fly, not the pitch of their honk. Geese fly best at a certain air density (ie pressure), which changes with the weather. In a depression, where there is low pressure, the geese fly low as the optimal air density for flying is lower. In an anticyclone, which normally brings fair weather, the air pressure is greater so the geese fly higher to find that optimal air density. However, practically this is extremely inaccurate, so I wouldn't recommend using it for actual weather forecasting.

When your joints all start to ache rainy weather is at stake

Here is another pressure-related proverb. When there is low pressure, it allows tissues to expand in the body which can put pressure on the joints, causing pain. This is why arthritic people often say that they can predict when bad weather is coming. Obviously, this proverb is not necessarily the most accurate as your joints could ache for countless other reasons too.

> **If the spiders are many and spinning their webs the spell will soon be very dry**

If the spiders are many and spinning their webs the spell will soon be very dry

In high humidity spiders' webs absorb water and become heavy, causing them to be more obvious to prey and prone to break. Spiders are very sensitive to moisture in the air and they are aware that their webs may break, so when there is high humidity they will hide and not spin their webs. When the spiders sense low humidity, they will come out and resume their web spinning. This is a useful indicator because if there are lots of spiders spinning their webs we know that fair weather is probably here for at least a day.

> **Frogs croaking in the lagoon means rain will come soon**

This is another proverb about humidity. Frogs need moist and warm skin to be active, and when they croak it means that the conditions are right, it is warm and there is high humidity. High humidity indicates poor weather, so beware of frogs croaking.

> **Seagull, seagull, stay out from the land we'll ne'er have good weather while you're on the sand**

Seabirds and wildfowl react to bad weather signs long before we may know about them. Seagulls which normally forage far out to sea will come close inshore to fish and will be noisy and fly carelessly and aimlessly when bad weather is coming.

The exception to this, though, is on beaches in the summer which will be busier when the weather is good and will attract more seagulls to steal from picnics.

No weather is ill, if the wind be still

Lack of wind is usually characterised by anticyclones, which normally bring fair weather and little cloud formation. The problem with this is that there is often no wind at night when a temperature inversion forms, but if there is a gradient wind the wind should have picked up by mid-morning. Another issue is that the wind may be light if there is a sea breeze, although generally if the wind is light during the day for more than an hour, that indicates the presence of an anticyclone.

When the wind backs and the barometer falls, then be on your guard against gales and squalls

As we have seen earlier, one of the first signs of a depression is a backing wind along with falling pressure. A depression will often bring gale force winds and squalls as the cold front passes. The photo below is a rather good example of a line squall.

When the bubbles of coffee collect in the centre of the cup, expect fair weather
When they adhere to the edges of the cup, forming a ring, expect rain

This weather proverb has been known since the 19th century and there seems to be some reasoning behind it. One theory is that high pressure pushes down on the surface of the coffee, making it slightly concave, causing the bubbles to move to the middle of the cup. Conversely when the pressure is low it makes the surface slightly convex so the bubbles move towards the edges of the cup. However, this proverb isn't particularly accurate – some people profess the exact opposite.

If we fog we don't frost and if we frost we don't fog

This is not really weather lore but a rule of thumb used by weather forecasters. Basically fog acts like a blanket over the land and it insulates the ground, stopping the heat from escaping. Equally, if there is a frost it means that the moisture used to make fog is trapped on the ground, making it unlikely to be be foggy. So if we see frost settling during a winter evening we know that there won't be fog and if we see fog during the night we know there won't be frost.

Mountains in the morning fountains in the evening

Tall cumulus clouds tower up high into the sky like mountains, and if these are seen early in the morning it indicates that the air is quite humid. These cumulus clouds then grow further during the day and by the afternoon it is likely to be showery, although these showers will tail off by the evening.

When dew is on the grass rain will never come to pass

Dew forms when the ground radiates out its heat at night and cools the air above it until the air can't hold any more water, so it sheds the water. This water condenses around the tiny particles on grass and other objects. But for dew to happen there needs to be a clear sky as this allows the earth to radiate heat away much better. These cloudless nights generally indicate anticyclonic conditions and it could be fair weather for the next few days.

WHERE DIFFERENT CLOUDS ARE FOUND ON A TYPICAL WEATHER CHART

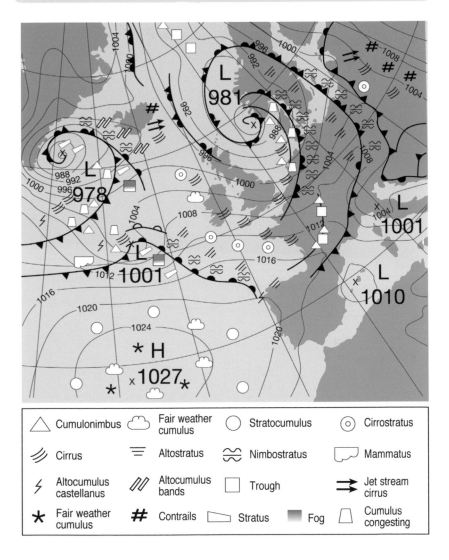

△	Cumulonimbus	☁	Fair weather cumulus	◯	Stratocumulus	◉	Cirrostratus
⫽⫽	Cirrus	═	Altostratus	≋	Nimbostratus	�	Mammatus
⚡	Altocumulus castellanus	⫽⫽⫽	Altocumulus bands	▢	Trough	⇒	Jet stream cirrus
✳	Fair weather cumulus	#	Contrails	▱	Stratus	▮ Fog	⬡	Cumulus congesting

QUIZ – WHAT DID YOU LEARN?

9

1. Which type of clouds are these? What weather do they indicate and how long might it be until that weather arrives?

2. These clouds were spotted at 11am. Which type of clouds are they and what will the weather be like for the rest of the day?

3. This cloud is upwind of you. Which type of cloud is it and what are the expected conditions under it?

4. Which type is this rather dull-looking cloud? What weather does it indicate and how long might it be until that weather arrives?

5. Which type of cloud is this? What weather does it indicate and how long might it be until that weather arrives?

6. You notice a lack of clouds over the sea but numerous cumulus clouds over the land. What effect does this indicate? What are the expected conditions for the next day after seeing this cloud, and why is this?

7. It is below freezing and you can see nimbostratus cloud upwind. Will there certainly be snow or might there be rain?

Answers on page 128

APPENDIX 1: LOGGING YOUR OBSERVATIONS

There is more to forecasting the weather than just looking at the clouds. It is important to record a few pieces of information to make a forecast more accurate. Record your observations a few times per day, more if the weather is changing rapidly.

You should observe and record:

1. Clouds

Write down not only what type of clouds are present but also the upper wind direction and speed ('fast' or 'slow' would do for this). Also work out the proportion of cloud cover.

2. Air

Make sure you record not only the reading but also the pressure changes. It is sensible to record this hourly.

3. Wind

Make sure there are no obstructions that may affect the wind direction. Throw some grass in the air and use a compass, or a wind indicator if you have access to one.

4. Wind speed

Use the Beaufort Scale to judge the speed (see page 115).

5. Visibility

Use general terms like 'good' or 'poor'.

6. Sea state

This can often warn of approaching depressions.

Also record any changes or events as and when they happen, such as rain showers or when you first notice a halo. If you have access to the right equipment recording the temperature or humidity can also be useful, especially for working out whether you are in a depression.

Now try to find out why you are experiencing these conditions. Remember to think about whether there would be any local effects that might impact the weather, such as sea breezes or mountains, which often drastically change the wind direction. Then try to think about what large-scale weather could cause these conditions: is there a large anticyclone giving you sunny and windless days, or is there a series of low pressure systems coming through with a few hours of fair weather before the next front starts to arrive?

If the wind seems to be backing and increasing in strength, the barometer is dropping and there are a few cirrus clouds about you may well infer that a depression is on its way. Equally, if you notice a few cirrus clouds in the sky but no drop in pressure, increase in wind strength and the upper wind is parallel to the lower wind, then you might predict that there isn't actually a depression coming.

APPENDIX 2: THE BEAUFORT SCALE OF WIND SPEED

Force	Wind speed (knots)	Name	Appearance
0	Less than 1	Calm	Smoke rises vertically, leaves don't move, sea like a mirror
1	1–3	Light air	Smoke drifts, wind vanes don't move, scaly ripples without foam crests
2	4–6	Light breeze	Wind felt on face, vanes move, leaves rustle, flags not extended, small wavelets with glassy crests
3	7–10	Gentle breeze	Light flags extended, leaves in constant motion, large wavelets, glassy crests now break
4	11-16	Moderate breeze	Most flags extended, small branches move, dust and small pieces of paper move, small waves with white horses
5	17-21	Fresh breeze	Small trees sway, moderate waves, many white horses
6	22–27	Strong breeze	Large branches move, wires whistle, some blown spray
7	28–33	Near gale	Whole trees move, difficulty walking upwind, foam blown in streaks along wind

Force	Wind speed (knots)	Name	Appearance
8	34–40	Gale	Rare inland but twigs break off trees, moderately high waves, obvious foam streaks
9	41–47	Strong gale	Slates and chimney pots fall, fences blown down, high waves, crests begin to topple, spray may affect visibility
10	48–55	Storm	Very rare inland, trees uprooted, structural damage, very high waves, sea surface looks white
11	56–63	Violent storm	Widespread damage, exceptionally high waves, sea completely covered in foam, wave crests blown into froth
12	63+	Hurricane	Devastation, phenomenally high waves, air filled with foam and spray

This was devised 200 years ago by Admiral Sir Francis Beaufort, who came up with a system to compare the wind speed to the amount of sail warships should have up. Even today it is relevant and extremely useful, as it is often the only way to estimate the wind's strength.

WHAT THE BEAUFORT SCALE MEANS FOR YOU

Be aware that every time the actual wind strength doubles the effect it has on an object is four times larger. This is because the force exerted by the wind is proportional to the square of the wind strength. But if the wind increases from 5 knots to 20 knots the force of the wind will actually be 16 times larger, so even a small increase in wind strength could be quite dangerous.

Force	What it means for sailors	For dinghy sailors	For walkers
0	Sails won't fill and boat drifts about	Telltales won't move	You are probably in a temperature inversion, so fog will not be seen on higher ground and a cloud inversion is likely
1	Sails just fill but boat still drifts. Spinnakers don't fill	Sails just fill and telltales fly, crews sit on opposite sides	Negligible wind chill and probably a temperature inversion like above, upslope fog possible
2	Boat steers well now and spinnakers generally fill. Wind is felt on your cheek	Crews sit on the same side now and burgees respond	0°C will feel like -3°C and -10°C like -15°C, upslope fog likely with the right wind direction and humidity
3	Boat moving well, light ensigns are extended	Crews sit up on the gunwale	Slightly more wind chill experienced, upper limit of upslope fog

Force	What it means for sailors	For dinghy sailors	For walkers
4	Upper limit of all sails being up	Crews lean out	0°C will feel like -6°C and -10°C like -19°C, upslope fog becomes low stratus clouds instead
5	Boat heels too much with full sail, so smaller jibs and mains carried but yachts at max speed. Spinnakers still up	Crews fully hiking, main eased in gusts. Occasional capsizes	0°C will feel like -7°C and -10°C like -20°C, fairly breezy, but isn't going to impact walking
6	All yachts now reefed, cruisers start to shelter	Dinghies are overpowered and there are many capsizes; it is difficult to keep the boat flat even when spilling wind	Umbrellas are hard to use and people stop going outside due to the wind, similar wind chill to above
7	Racing yachts may still carry spinnakers but most sailors seek shelter	Most racing called off, experienced sailors enjoy reaching about	You will have difficulty walking but are still probably safe from branches falling, 0°C will feel like -9°C and -10°C like -22°C
8	Storm sails up or boats heave to; most sailors won't experience windier conditions than this	Dinghies only sailed by experienced sailors, races abandoned mid race in these conditions	You will have to lean into the wind to walk upwind and will be slightly blown about, forested areas become hazardous due to falling branches, similar wind chill to above

ABOUT THE AUTHOR

Oliver Perkins is an author and freelance journalist who has written for magazines including *Yachting Monthly*, *Practical Boat Owner* and *Country Living*. He used to sail for the British Youth Sailing team and has sailed his family yacht out of Bosham for as long as he can remember. He lives in Guildford and also enjoys climbing. He has also written the *Reeds Cloud Handbook* for Adlard Coles.

ACKNOWLEDGEMENTS

Thanks especially to my parents, Chris and Hilary, who ignited my interest in sailing and for putting up with my persistent weather forecasting. And thanks to my brother Ben who helped me take some of the photos.

Thanks to Tom Cunliffe and Duncan Wells for so kindly writing the Forewords and to Professor Christopher Collier for fact checking the book.

Finally thank you to Eleo Carson, who extremely generously edited the original version of this book for me.

PHOTO CREDITS

All photos are the property of the author with the exception of: Anton Yankovyi (126 mammatus), Bidgee (39), Bieverwin (32), contri (59), F A Martin (126 stratus), GerritR (43 top, 55), Getty (4, 22, 23, 24 top and bottom, 25 top, 31, 33, 34, 36 top and bottom, 40, 31, 42 top and bottom, 45, 49, 50, 53, 57, 58, 60, 61, 62, 69, 72, 74, 85, 94, 96, 103, 104, 105, 106, 124 all except altostratus, 126 all except pannus, stratus and mammatus), Jmcc150 (47), Simon Eugster (126 pannus), Wikimedia (124 altostratus)

BIBLIOGRAPHY

Bartlett, Tim, *Weather Companion*: Fernhurst, 1999.

Dunlop, Storm, *Weather: Collins Nature Guide*: Collins, 2012.

Houghton, David, *Wind Strategy*: Fernhurst Limited, 2016.

Pike, Dag, *50 Ways to Improve Your Weather Forecasting*: Adlard Coles, 2007.

Shonk, Jon, *Introducing Meteorology*: Dunedin, 2013.

Watts, Alan, *Instant Weather Forecasting*: Adlard Coles, 2011.

Watts, Alan, *The Weather Handbook*: Adlard Coles, 2014.

Westwell, Ian, *Weather*: PRC, 1999.

Cunliffe, Tom, *200 Skippers Tips*: Wiley Nautical, 2010.

Gooley, Tristan, and Neil Gower, *How to Read Water*: Sceptre, 2017.

Houghton, David, *Weather at Sea*: John Wiley & Sons Ltd, 2008.

Karnetzki, Dieter, *This is Practical Weather Forecasting*: A & C Black, 1994.

Keeling, Simon, *The Sailor's Book of the Weather*: John Wiley & Sons, 2008.

GLOSSARY

Air masses Bodies of air in which the temperature and humidity is constant.

Air pressure The weight of the air above a particular point.

Anticyclones Areas of higher air pressure than the surrounding area.

Back A wind shift anticlockwise.

Depressions An area where the pressure is lower than the surrounding area.

Diffraction Bending of waves around an obstacle.

Eddy An eddy is a reverse current in the air or water caused by the wind or current reaching an obstruction.

Fetch The length of water over which the wind has blown.

Front A narrow zone that separates cold and warm air masses.

Gradient wind Lower wind without local wind effects.

Jet streams Fast flowing, narrow, meandering air currents found just below the tropopause in the atmosphere. They flow around the earth from west to east.

Multi-cell thunderstorms A cluster of thunderstorms with cells at different stages in the life cycle of a thunderstorm.

Occluded front A front where the cold front catches up with the warm front and forces the warm sector above the ground.

Sea breeze A wind that blows from the sea to the land due to the higher temperature of the land than the sea.

Swell waves Waves which were created somewhere else and moved in to the location.

Temperature inversion A layer in which the temperature increases with height.

Thermally-enhanced wind Wind strengthened due to compression isobars caused by a warming of the land.

Tropopause The top boundary of the troposphere. Above this the temperature increases with height.

Troposphere The lowest layer of the atmosphere, which is around 10km thick.

Trough An extended region of relatively low atmospheric pressure. It is basically a valley in an air mass which clouds pool up in.

Veer A clockwise wind shift.

Virga Also known as fallstreaks, virga are trails of descending ice crystals or water droplets that evaporate before they reach the ground.

Warm sector The area of warm air found between a cold front and a warm front.

Waterspout A rotating column of water caused by a tornado forming over a body of water.

Wind bands Long strips of areas of stronger wind and lighter wind that are orientated in the direction of the wind.

Wind shear Wind shear is a difference in wind speed or direction over a relatively short distance in the atmosphere.

INDEX

	Cloud name	What weather it indicates
	Cirrus	Rain, wind and possibly frontal fog (depression approaching)
	Hooked cirrus	Lots of rain and wind (vigorous depression approaching)
	Jet stream cirrus	Winds increase significantly (vigorous depression approaching)
	Fair weather cirrus	Light winds and no precipitation
	Cirrostratus	Rain, wind and possibly frontal fog (depression approaching)
	Cirrocumulus	Rain, wind and possibly frontal fog (depression approaching)
	Cold front passed	Improvement, with lighter winds and less rain although there may be a few heavy showers
	Abnormally thick contrails	Rain, wind and possibly frontal fog (depression approaching)
	Altostratus	Rain, wind and possibly frontal fog (depression approaching)
	Altocumulus before a warm front	Rain, wind and possibly frontal fog (depression approaching)
	Thundery altocumulus	Showery conditions and possibly thunderstorms
	Fair weather cumulus	If seen after mid-morning fair weather

When	Things to be aware of
12 24 hours	If they aren't increasing in number they may instead indicate fair weather
12–24 hours	True hooked cirrus clouds almost always indicate fronts are approaching
6–12 hours	You usually see the clouds moving across the sky
Probably at least a day	If these start to rapidly increase in number be careful as a front may be on its way!
10–15 hours	These won't always form a halo
8–12 hours	This is often not true as they may indicate thunderstorms or even fair weather!
1–2 hours	Use the cross-winds rule to check whether another front is coming
12–24 hours	Always look out for other clouds to confirm if a front is coming
Less than 8 hours	This is only the case if the cloud base is lowering
Less than 8 hours	These can form at other times and are often confused with the lower stratocumulus
Within the next 12 hours	Remember ragged or bulging altocumulus clouds indicate thunderstorms
For the rest of the day	These are only the ones less tall than they are wide

	Cloud name	What weather it indicates
	Coastal cumulus	Windier and showery conditions out to sea
	Cumulus congestus	Showers, gusts and possibly thunder
	Trough	Windy and showery conditions
	Pannus	Rain
	Stratus	Stable cloudy conditions with a chance of drizzle
	Stratocumulus	Dry and stable cloudy conditions
	Nimbostratus	Mainly continuous rain (not showers)
	Mammatus	Storm, due to cumulonimbus clouds
	Cumulonimbus	Showers, gusts and possibly thunder
	Thunderstorm	Thunder, lightning, showers and gusts

When	Things to be aware of
Now	The wind will probably be a couple of Beaufort forces stronger
Within 3 hours	If these are first seen in the afternoon showers are unlikely
For around an hour	These are different from a front as the weather goes back to what it was before the trough arrived
In a few minutes	They can be seen under all types of rain clouds
At least the next few hours	They often produce drizzle when going over hills or mountains
At least for the next few hours	These can take many different forms, so any low cloud that isn't cumulus or stratus is considered stratocumulus
Usually for no more than 4 hours	These are almost exclusively found in depressions, so expect very changeable conditions
In a few minutes (or it could have just happened or might be happening now)	They can also form under other non-rain bearing clouds, so be careful interpreting these ones
In a few minutes if these are upwind of you	They are usually isolated so although you might see these you could get lucky and they might all miss you
In a few minutes	They can last for any length of time, from about quarter or half an hour to a few hours

QUIZ ANSWERS

1. Cirrus clouds indicate prolonged rain in a warm front, which will arrive in 12–20 hours.

2. Cumulus humilis clouds seen after mid-morning indicate fair weather for the rest of the day.

3. Beware – it is a cumulonimbus cloud and you should expect heavy rain, probably hail and winds of up to 40 knots.

4. Pannus clouds indicate that it is already raining, but the rain is evaporating before it reaches you. But expect rain in less than 15 minutes.

5. Halos form in cirrostratus clouds and they indicate prolonged rain in a warm front, which will arrive in 10–15 hours.

6. The line of clouds over the coast is a sea breeze front, where the air is rising, and the lack of clouds over the sea is due to the falling air over the sea. This indicates that there is a sea breeze. The sea breeze there would be quite light as the photo was taken at the end of March, at 4pm, and the sea breeze front is only a couple of kilometres inland.

7. If the snow has passed through a layer of slightly warmer air higher up, this can melt the snow, which then falls as freezing rain, so there won't certainly be snow.